Imitation Learning for Robots: Building a Strong Foundation

Jacob

Copyright © 2024 by Jacob

Contents

2. Introduction

Industrial robotics is contributing to the largest world's economies. To mention an example, in 2016, sales of industrial robots in China reached 87.000 units, accounting for around 30 percent of the global market (Cheng et al., 2019). Robotics provides tools for automating industries (Figure 2.1), making the production chain faster and more reliable. The current advantage of robots over human being workers is their ability to effortlessly replicate with high precision a task thousands of times per day. However, a substantial downside of current robotics is the poor adaptation to new tasks or situations.

Take, as an example, a company that wants to change its product. In this case, the company needs to reconfigure part of the production line and reconfigure some robots. The re-configuration of the robots requires high-qualified engineers and is, in general, expensive. Recent progresses in machine learning and artificial intelligence hold the promise of simplifying the re-configuration process via robot learning.

Figure 2.1.: An example of the integration of robots in a industrial environment. Enhancing their pre-programmed behavior with robot-learning provides better adaptability[1].

However, this tremendous progress is mainly limited to simulated tasks, such as video-games, board-games, and simulated environments (Mnih et al., 2015; Schulman et al., 2015, 2017; Silver et al., 2017, 2018).

We identify two primary sources of this problem: (i) lack of sample efficiency and, (ii) lack of safety guarantees. While simulated tasks can be parallelized and accelerated w.r.t. reality, robotic systems are slow, requiring the learning system to learn with fewer interactions (samples). Current state-of-the-art techniques often require millions of samples. Furthermore, learning usually needs errors to learn. However, lousy behavior might break expensive robotic systems or even become harmful to human beings. For this reason, we need to guarantee that the robot will learn and behave safely. However, such guarantees are usually hard to be delivered. In this book, we focus on mitigating these two problems.

Robot-learning is usually composed of two phases: in a first imitation learning phase the robot aims to replicate some expert behavior; and a second reinforcement-learning phase, where the robot improves its behavior (Schaal, 1999; Argall et al., 2009; Muhlig et al., 2009). The imitation-learning phase is usually crucial in robotic applications, as it aims to initialize the robot's behavior with a sound and reliable policy. However, imitation-learning usually has a poor outcome, as the learning system aims to imitate the expert without knowing the demonstration's goal. In the subsequent reinforcement-learning phase, the robot has to interact with the system, and a user provides a reward signal dependent on the interaction's outcome. A reinforcement-learning algorithm has to refine the robot's behavior to maximize the reward signal (Sutton & Barto, 2018).

---

[1] Granted by `RGRobotics.com` under Creative Common License

In this book, we aim to mitigate the mentioned issues (sample inefficiency and lack of safety guarantees) by analyzing and improving state-of-the-art imitation learning and reinforcement learning. This chapter gives a brief overview of imitation learning and reinforcement learning, focusing on state-of-the-art techniques. We terminate this chapter by summarizing our contributions.

One core idea of this book is to extrapolate a low dimensional representation of the robot's behavior during the imitation learning phase via dimensionality reduction (Lee & Verleysen, 2007). This low-dimensional projection has the twofold effect of making the subsequent reinforcement learning both safer and more sample efficient (Colomé et al., 2014; Colomé & Torras, 2018a). Chapter 3 analyzes an efficient low-representation of the robotic's behavior (a.k.a. policy). In Chapter 4, we see how to combine the low-dimensional policy's representation with reinforcement learning (Tosatto et al., 2020d).

Another contribution of this book is to develop a particular kind of sample-efficient reinforcement learning. As we will see later, one of the significant sources of inefficiency in reinforcement learning is the wide adoption of on-policy techniques. Off-policy techniques hold the promise of delivering higher efficiency and more robust safety, however, the off-policy estimation is usually more complex. In Chapter 3, we introduce an off-policy estimation based on pre-existing techniques (Levine et al., 2020) that deliver exceptionally high efficiency.

However, the kind of reinforcement learning introduced in Chapter 4 is not suited to complex multi-step scenarios (where the robot has to act and observe the scene multiple times before reaching task completion). The off-policy estimation is, in fact, harder for multi-step problems. In Chapter 5, we present a novel off-policy technique that is well suited to multi-step problems. We show empirically that our techniques can achieve optimal behavior from a few poorly-performing demonstrations. The presented technique is accompanied by statistical guarantees (Tosatto et al., 2020b,c). In Chapter 6, we analyze the statistical guarantee provided in Chapter 5 (Tosatto et al., 2020a). A generic problem of reinforcement learning is the so-called exploration/exploitation dilemma (Carpentier et al., 2011). This dilemma becomes complex in the more common multi-step scenario. Chapter 7 proposes a novel technique based on an optimistic estimator, which can deliver a better exploration (Tosatto et al., 2019). To conclude, Chapter 8 summarizes the core contributions, analyzes our inspections' findings, and outlines future research directions.

All the developed methods are tested with numerical simulations, and some are tested on real robotic setups. Figure 2.6 depicts the real robotic platform used within an example of the phases involved. In Chapters 3 and 4 we tested out techniques on a pouring task, involving state observation (i.e., the weight in the glass), human demonstrations and policy improvement. This is an example of the task which resembles a challenging manipulation task, where the robot must behave accordingly to new situations (e.g., it should pour a precedently unseen amount in the glass), and must deal with the presence of noisy observations.

2.1. Motivations

Learning in a real system provides many challenges. Robotic movements are encoded with a large number of parameters. The learning process must explore safely (to not damage hardware or harm people) and efficiently. Two keys contributions of this book are the usage of dimensionality reduction techniques and off-policy reinforcement learning. In the following, we detail our motivations mentioning state-of-the-art techniques with their advantages and their disadvantage, highlighting their usage in our work.

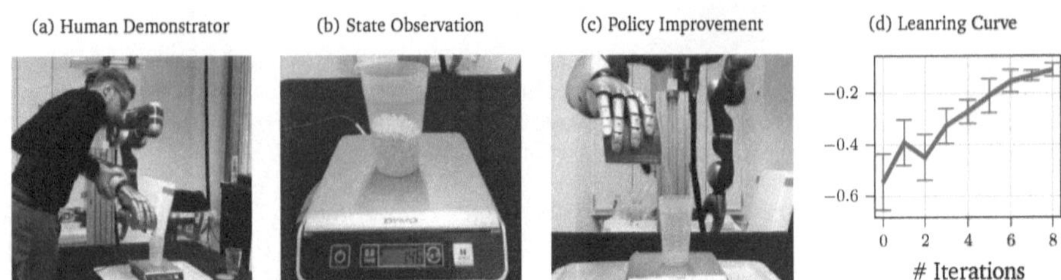

Figure 2.2.: The pouring task resembles an industrial process. We provide a dataset of demonstrations on how to reach a desired amount of liquid poured in the glass (a). The movement is *conditioned* by the desired amount. The current state is acquired by a device, in this case a scale (b). The robot interacts with the system (c) and improves its behavior (d).

2.1.1. Why Dimensionality Reduction

Suppose to observe 100×100 images of a ball in different 3d positions with white background. In this scenario, it exists a deterministic mapping $g : [0, 1]^3 \to [0, 1]^{10000}$ mapping the ball's position into the image, which can be seen as a 10000 vector of values between 0s and 1s. Unfortunately, we cannot observe the original 3d positions of the ball, but only the 100×100 images. Let us now consider, that uniformly sampling an image in $[0, 1]^{10000}$, will almost surely results in a image that does not represent a ball, as the co-domain of g occupies only a small portion of $[0, 1]^{10000}$. As Figure 2.3 depicts, sampling a 3-d point in $[0, 1]^3$, will *always* maps to a "valid" image of a ball in $[0, 1]^{10000}$. This naïve example suggests that finding g and g^{-1} will allow (i) to safely generate valid samples in the smaller space (properly called *latent space*) and (ii) since the latent space is smaller than the observation space, we can usually perform optimizations more efficiently. In Chapters 3 and 4, the policy is encoded as a movement primitive (Paraschos et al., 2013, 2018). In short, the robot's movement is described as a linear combination between a set of features and a set of radial

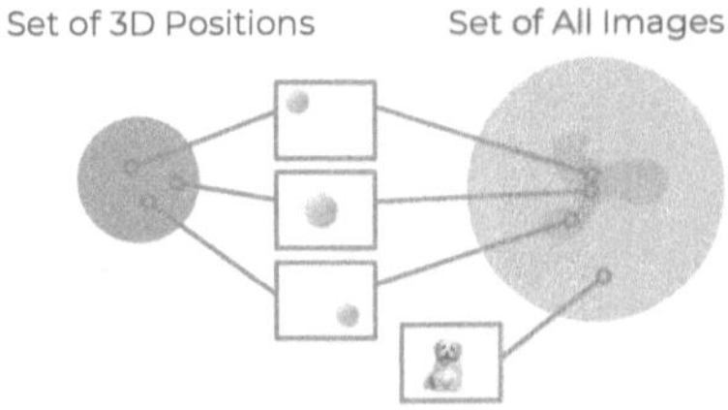

Figure 2.3.: Knowing a mapping between 3d positions and images allows us to always generate an image of a ball given any 3d position. On the contrary, the set of images is far larger and includes non-related pictures. For this motivation we argue that working in the appropriate lower dimensional space is convenient and safer in robotic applications.

basis functions. To represent complex articulated movement, one needs to use many parameters (typically about a hundred). When the reinforcement learning algorithm optimizes the parameters, it explores in the space of all the possible movements representable by the mentioned linear combination. This exploration is inefficient and unsafe. We propose to use dimensionality reduction on a set of human demonstration to find a better exploration space. To this end, in Chapter 3, we use the principal component analysis (PCA) on the movement's parameters. The mixture of probabilistic component analysis, used in Chapter 4 overcome some of PCA issues.

Principal component analysis (PCA) is perhaps the most well known dimensionality reduction technique (Pearson, 1901; Jolliffe, 1986; Bishop, 2006). PCA can have different interpretations. In this section, we follow the derivation proposed by Hotelling (1933), which analyzes the problem in terms of variance maximization of a linear projection. Consider a set of observations $\{y_i\}_{i=1}^N$ where $y \in \mathbb{R}^d$. Suppose, for the moment, that our objective is to project the variables y into a one-dimensional space $\mathbb{R}^{d_l}$ where $d_l = 1$. Define $\bar{y}$ as the empirical mean and Σ the empirical covariance of $\mathbf{y}$. Let us consider a unit projecting vector $\mathbf{u}$ (where $\mathbf{u}^\top\mathbf{u} = 1$). We notice that the variance of the projecting points is given by $\sigma_z^2 = \mathbf{u}^\top\Sigma\mathbf{u}$. We now maximize the variance σ_z^2 w.r.t. $\mathbf{u}$. Using Lagrange multiplier, we can solve this optimization problem by imposing the Lagrangian

$$\mathcal{L} = \mathbf{u}^\top\Sigma\mathbf{u} + \lambda(1 - \mathbf{u}^\top\mathbf{u}) \tag{2.1}$$

to have zero gradient

$$\nabla_\mathbf{u}\mathcal{L} = 0 \implies \Sigma\mathbf{u} = \lambda\mathbf{u}. \tag{2.2}$$

Equation 2.2 states that $\mathbf{u}$ must be an eigenvector of Σ and λ must be the correspondent eigenvalue. By multiplying (2.2) on both the side by $\mathbf{u}^\top$, we notice that $\lambda = \sigma_z^2$. Therefore, to obtain the highest possible variance, we must choose the highest eigenvalue (with its correspondent eigenvector). The principle behind the fact that we want to maximize the variance's projection, can be seen as if we want to keep the maximum amount of information. To extend this derivation to a multivariate projection, one can iterate the exposed process to obtain a matrix of eigenvectors $\mathbf{U}$. PCA is not based on a probability model. This deficiency leads to the impossibility of probabilistic conditioning. Probabilistic conditioning is essential in our robotic tasks, as it allows to sample a movement based on a particular condition (e.g., on a desired weight in a pouring task, or on a object position for a reaching task). This handicap is overcome with the treatment of PCA as a probabilistic model.

Probabilistic principal component analysis allows probabilistic conditioning, noise cancellation and density estimation at once. Suppose that the observed data is generated via a linear relation w.r.t. the latent space and perturbed with an isotropic Gaussian noise, i.e.,

$$\mathbf{y} = \mathbf{Wx} + \mu + \epsilon, \qquad \epsilon \sim \mathcal{N}(\epsilon|0, \sigma^2\mathbf{I}), \tag{2.3}$$

hence $\mathbf{y}|\mathbf{x} \sim \mathcal{N}(\mathbf{y}|\mathbf{Wx} + \mu, \sigma^2\mathbf{I})$. Conveniently, we define the marginal distribution of the latent variable $\mathbf{x} \sim \mathcal{N}(\mathbf{x}|0, \mathbf{I})$. The marginal distribution of $\mathbf{y}$ under this assumption becomes $\mathbf{x} \sim \mathcal{N}(\mathbf{x}|\mu, \mathbf{WW}^\top + \sigma^2\mathbf{I})$. The log-likelihood of the joint probability of the observations takes the form of

$$\mathcal{L} = -\frac{N}{2}\left(d\log(2\pi) + \ln|\mathbf{WW}^\top + \sigma^2\mathbf{I}| + \mathrm{tr}\left((\mathbf{WW}^\top + \sigma^2\mathbf{I})^{-1}\Sigma\right)\right) \tag{2.4}$$

where Σ is the empirical covariance of the observations. The maximization of (2.4) w.r.t. $\mathbf{W}$ and σ^2 can be obtained in closed-form,

$$\mathbf{W}_{\mathrm{ML}} = \mathbf{U}_{d_l}\left(\Lambda_{l_q} - \sigma^2\mathbf{I}\right)^{1/2}\mathbf{R}, \tag{2.5}$$

where $\mathbf{U}_{d_l}$ is the matrix of the eigenvectors of Σ corresponding to the highest d_l eigenvalues, Λ_{d_l} is a diagonal matrix containing the eigenvalues on its diagonal, and $\mathbf{R}$ is an arbitrary orthogonal rotation matrix. The maximum-likelihood solution of σ^2 is

$$\sigma_{\mathrm{ML}}^2 = \frac{1}{d - d_l}\sum_{i=d_l+1}^{d}\lambda_i.$$

For further details, refer to Tipping & Bishop (1999b).

PCA and PPCA rely on linear projection, exhibiting unsatisfactory performance when the dataset contains multimodalities. This often the case of robotic movements. To mitigate this problem, Tipping & Bishop (1999a) formulated a probabilistic model composed of a **mixture of probabilistic principal component analyzers (MPPCA)**. MPPCA can be used to perform density estimation, sample generation, sample reconstruction, noise-cancellation, and perform satisfactorily on a fair number of dimensions and complex dataset, such as the MNIST. Figure 2.4 depicts a reconstruction of images present in the MNIST dataset. As can be seen MPPCA effectively reconstructs 28×28 pixel images.

In MPPCA ones define a number of clusters K. A categorical distribution parametrized with π defines the probability of selecting a specific cluster. Each cluster k encodes a linear relation between a standard-distributed latent space $\mathbf{x}$ and the observation space. Similarly to PPCA, we provide an additional isotropic noise ϵ with different magnitudes σ_k for each cluster. The model described can be summarized as follows,

$$k \sim Cat(k|\pi) \qquad (k \in \{1, \ldots K\})$$
$$\mathbf{x} \sim \mathcal{N}(\mathbf{x}|\mathbf{0}, \mathbf{I}) \qquad (\mathbf{x} \in \mathbb{R}^{d_l})$$
$$\epsilon \sim \mathcal{N}(\epsilon|\mathbf{0}, \mathbf{I})$$
$$\mathbf{y} = \mathbf{W}_k \mathbf{x} + \overline{\mathbf{y}}_k + \sigma_k^2 \epsilon. \qquad (\omega \in \mathbb{R}^m)$$

Via classical expectation-maximization, we can maximize the log-likelihood of $p(\mathbf{y})$ w.r.t. $\mathbf{W}_k, \overline{\mathbf{y}}_k, \sigma_k^2$. For further details, refer to Tipping & Bishop (1999a). An efficient implementation in python can be found at `https://github.com/SamuelePolimi/MPPCA`.

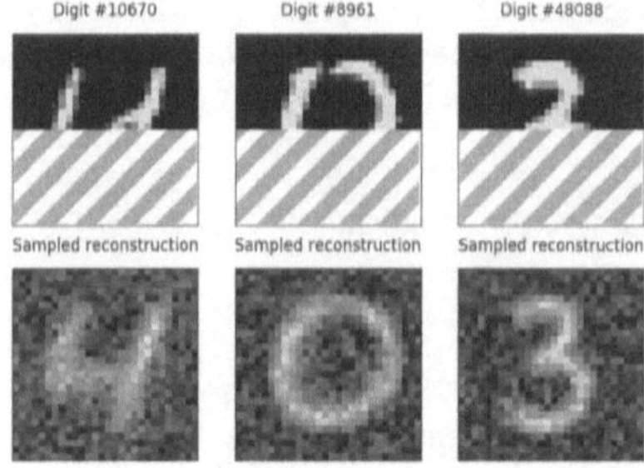

Figure 2.4.: An example of MPPCA on high dimension. In this example, MPPCA succesfully reconstructs MNIST digits (784 dimensions) from partial information.

To conclude, dimensionality reduction is a powerful tool that allows us to operate in a more convenient, space. In particular, MPPCA offers a simple probabilistic model that enables a principled treatment of multimodalities of the data and full mathematical tractability for probabilistic conditioning and noise cancellation. MPPCA can be applied efficiently to almost a thousands dimensions, ad depicted in Figure 2.4. The properties of MPPCA make it as an ideal candidate in our robotic applications.

2.1.2. Why Off-Policy Reinforcement Learning

Reinforcement learning aims to improve the behavior of an artificial agent by maximizing a perceived reward signal. The agent usually observes an environment, and at each time step, it will act accordingly. The agent's acting has an immediate impact on the future state of the environment and will impact all future reward signals. The agent's behavior is usually encoded by a set of parameters. Each reinforcement learning algorithm's core is to improve those parameters to obtain an expected higher cumulative reward.

The most common framework used to mathematically describe the reinforcement learning problem is the Markov decision process (MDP)[2].

An MDP $\mathcal{M}$ is described by the tuple $(\mathcal{S}, \mathcal{A}, \gamma, P, R, \mu_0)$ where $\mathcal{S}$ is the state space, $\mathcal{A}$ the action space, the discount factor $\gamma \in [0, 1)$. The transition probability from a state s to $s'e$ given an action a is governed by the conditional density $P(s'|s, a)$. The stochastic reward signal R for a transition $(s, a, s') \in \mathcal{S} \times \mathcal{A} \times \mathcal{S}$ is drawn from a distribution $R(s, a, s')$ with mean value $\mathbb{E}_{s'}[R(s, a, s')] = r(s, a)$. The initial distribution $\mu_0(s)$ denotes the probability of the state $s \in \mathcal{S}$ to be a starting state. A policy π_θ is a stochastic or deterministic mapping from $\mathcal{S}$ onto $\mathcal{A}$, usually parametrized by a set of parameters θ. We define an *episode* as $\tau \equiv \{s_t, a_t, r_t, \gamma_t\}_{t=0}^{T-1}$ where

$$s_0 \sim \mu_0(\cdot); \quad a_t \sim \pi(\cdot \mid s_t); \quad s_{t+1} \sim p(\cdot \mid s_t, a_t); \quad r_t \sim R(s_t, a_t, s_{t+1}).$$

---

[2]In Chapter 5, we use a generalized discount factor, which allows to elegantly define the task termination (White, 2017). However, to maintain the notation uncluttered, we use here the most classical notation with a constant discount factor.

The whole reinforcement learning is about the maximization of

$$\max_{\boldsymbol{\theta}} J(\boldsymbol{\theta}) = \max_{\boldsymbol{\theta}} \mathbb{E}\left[\sum_{t=0}^{T-1} \gamma^t r_t\right]. \tag{2.6}$$

When $T \leq \infty$ we can set $\gamma = 1$ and in this case we refer to "finite-horizon", while in the "discounted infinite horizon case" $T \to \infty$ and $0 \leq \gamma < 1$. In the averaged case, $\lim_{T\infty} T^{-1} \sum_{t=0}^{T-1} r_t$. In the following we will discuss how to improve the policy π_θ observing the set of samples. There are mainly two ways to solve the problem (2.6): episodic reinforcement learning based on parameter perturbations and step-based based on action perturbation.

Episodic. In episodic approaches, we treat the reinforcement learning problem as a classic optimization process without considering underlying state-action formalism. The idea is to set different values for the parameter $\boldsymbol{\theta}$ and observe the outcome by interacting with the environment and then updating $\boldsymbol{\theta}$ in the most promising direction. For this motivation, we call episodic reinforcement learning a black-box approach. the idea behind REINFORCE (Williams, 1992), one of the most well-known approaches, is to define a distribution $p(\boldsymbol{\theta}|\omega)$ over the policy's parameters $\boldsymbol{\theta}$, and to optimize the expectation

$$\max_{\omega} \mathbb{E}_{\theta \sim p(\boldsymbol{\theta}|\omega)} [R(\boldsymbol{\theta})] \tag{2.7}$$

where $R(\boldsymbol{\theta})$ denotes the total cumulative reward obtained with the parameters $\boldsymbol{\theta}$. To optimize the presented objective, we can proceed by estimating the gradient, and to perform a gradient-ascent update. The gradient can be estimated via samples,

$$\begin{aligned}
\nabla_\omega \mathbb{E}_{\theta \sim p(\boldsymbol{\theta}|\omega)} [R(\boldsymbol{\theta})] &= \nabla_\omega \int R(\boldsymbol{\theta}) p(\boldsymbol{\theta}|\omega)\, \mathrm{d}\boldsymbol{\theta} \\
&= \int R(\boldsymbol{\theta}) p(\boldsymbol{\theta}|\omega) \nabla_\omega \log p(\boldsymbol{\theta}|\omega)\, \mathrm{d}\boldsymbol{\theta} \\
&= \mathbb{E}_{\theta \sim p(\boldsymbol{\theta}|\omega)} [R(\boldsymbol{\theta}) \nabla_\omega \log p(\boldsymbol{\theta}|\boldsymbol{\omega})] \\
&\approx N^{-1} \sum_{i=1}^{N} R(\boldsymbol{\theta}_i) \nabla_\omega \log p(\boldsymbol{\theta}|\omega) \qquad \text{with } \boldsymbol{\theta}_i \sim p(\boldsymbol{\theta}_i). \tag{2.8}
\end{aligned}$$

Episodic reinforcement learning is often suitable to robotics, as (i) in many applications the policy parameters can be directly seen as an action (e.g., the parameters of a robotic movements), (ii) this framework allows easily to explore in parameter space keeping a deterministic policy π_θ to interact with the system and (iii) it works also in non-Markovian settings. A more sophisticated version of this gradient estimate has been used in Chapter 4.

Step-Based. The episodic view does not scale for large parametrization of the policy (e.g., deep neural networks). Instead, the step-based view uses the MDP assumption and proposes policy updates that depend on each visited state-action pairs. Two step based approaches are (i) policy-gradient based (like in Chapter 5) and (ii) dynamic programming approaches (like in Chapter 7). We propose an uncommon derivation of the policy gradient theorem that gives some insights on the method developed in Chapter 5. Let us consider a finite state space $\mathcal{S}$ and a finite action space $\mathcal{A}$ in the discounted infinite-horizon setting. At each state we consider the expected cumulative discounted reward $v_s = \mathbb{E}[\sum_{t=0}^{T-1} r_t | s_0 = s]$ and at each state-action pair

$q_{s,a} = \mathbb{E}[\sum_{t=0}^{T-1} r_t | s_0 = s, a_0 = a]$. The quantities v_s and $q_{s,a}$ satisfy Bellman equations

$$v_s = r_s^\pi + \gamma \sum_{s'} v_{s'} P^\pi(s'|s) \quad \text{with } r_s = \sum_a r_{s,a} \pi(a|s) \text{ and } P^\pi(s'|s) = \sum_a P(s'|s,a)\pi(a|s) \tag{2.9}$$

$$q_{s,a} = r_{s,a} + \gamma \sum_{s'} v_{s'} P(s'|s,a), \tag{2.10}$$

where $r_{s,a} = \mathbb{E}[R(s,a)]$. Similarly, we define a discounted stationary state-visit μ_s which encodes how many times, in average, the state s is visited starting from the state-distribution μ_0 and discounting by γ each visit. We can also establish a similar recursive relation,

$$\mu_s = \mu_0(s) + \gamma \sum_{s'} P(s|s')\mu_{s'}. \tag{2.11}$$

Let us combine in vector notation the value function $\mathbf{v}_\pi$, the state-dependent rewards $\mathbf{r}_\pi$, a transition matrix $\mathbf{P}_\pi$ defined as $P^\pi_{s,s'} = P^\pi(s'|s)$, the initial state distribution μ_0 and the state-visitation vector μ_π. Using the relations (2.9) and (2.11) we can solve in closed form for $\mathbf{v}_\pi$ and μ_π,

$$\mathbf{v}_\pi = \mathbf{r}_\pi + \gamma \mathbf{P}_\pi \mathbf{v}_\pi \implies \mathbf{v}_\pi = (\mathbf{I} - \gamma \mathbf{P}_\pi)^{-1} \mathbf{r}_\pi \tag{2.12}$$

$$\mu_\pi = \mu_0 + \gamma \mathbf{P}_\pi^\mathsf{T} \mathbf{v}_\pi \implies \mathbf{v}_\pi = (\mathbf{I} - \gamma \mathbf{P}_\pi)^{-\mathsf{T}} \mu_0. \tag{2.13}$$

The expected return defined can be conveniently rewritten as

$$J(\theta) = \mu_0^\mathsf{T} \mathbf{v}_\pi. \tag{2.14}$$

The gradient can be computed by substituting in (2.14) the closed-form solution of $\mathbf{v}_\pi$ in (2.12),

$$\begin{aligned}
\nabla_\theta J(\theta) &= \nabla_\theta \mu_0^\mathsf{T} (\mathbf{I} - \gamma \mathbf{P}_\pi)^{-1} \mathbf{r}_\pi \\
&= \mu_0^\mathsf{T} (\mathbf{I} - \gamma \mathbf{P}_\pi)^{-1} (\nabla_\theta \mathbf{r}_\pi + \gamma \nabla_\theta \mathbf{P}_\pi \mathbf{v}_\pi) \\
&= \mu_\pi^\mathsf{T} (\nabla_\theta \mathbf{r}_\pi + \gamma \nabla_\theta \mathbf{P}_\pi \mathbf{v}_\pi) \\
&= \sum_s \mu_s \left(\sum_a r_{s,a} \nabla_\theta \pi_\theta(a|s) + \gamma \sum_{s'} v_{s'} \sum_a P(s'|s,a) \nabla_\theta \pi(a|s) \right) \\
&= \sum_s \mu_s \sum_a \pi(a'|s) \left(r_{s,a} + \gamma \sum_{s'} v_{s'} P(s'|s,a) \right) \nabla_\theta \log \pi(a|s) \\
&= \sum_s \mu_s \sum_a \pi(a'|s) \nabla_\theta \log \pi(a|s) q_{s,a}
\end{aligned} \tag{2.15}$$

Equation 2.15 corresponds exactly to the policy gradient Theorem (Sutton et al., 2000). This derivation clearly highlights that the state-action visitation $\mu_s \pi(a|s)$ is generated by the application of the gradient to the closed-form solution of the Bellman equation. We use this principle to obtain an off-policy estimate of the policy gradient Theorem in Chapter 5. **On-Policy** Equation 2.15 gives us a powerful tool to estimate the policy-gradient via samples. The term $\mu_s \pi(a|s)$ denotes the discounted state-action visitation (or distribution, if normalized), and therefore

$$\nabla_\theta J(\theta) \approx \sum_{t=0}^{T-1} \gamma^t Q(s_t, a_t) \nabla_\theta \log \pi(a|s), \tag{2.16}$$

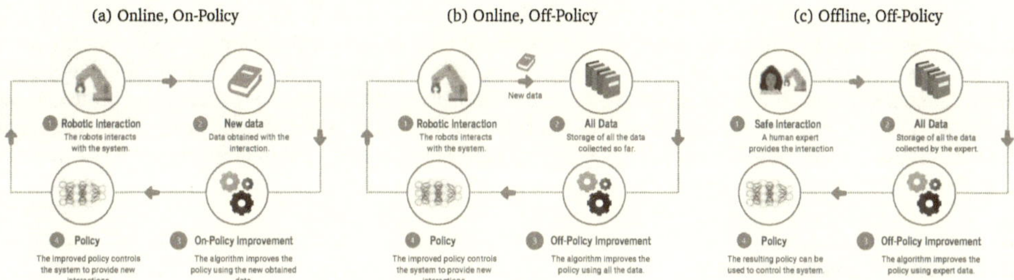

Figure 2.5.: In on-policy reinforcement learning (a), the policy improvement must rely only on the last data collected, making the process inefficient. In online, off-policy reinforcement learning, the policy improvement allows the usage of past data collected by the robot, increasing data efficiency. In offline, off-policy reinforcement learning (c), the policy improvement works on an external source of data, like a human demonstrator. This allows the system to improve the policy without interacting on the system providing both safety and efficiency.

where $Q(s_t, a_t) = \mathbb{E}[\sum_{i=t}^{\infty} \gamma^{i-t} r_i]$ can be estimated via Monte-Carlo sampling or via approximate dynamic programming[3]. Equation 2.16, however, forces one to estimate the gradient of π_θ by interacting with the environment using π_θ: in principle, one cannot use other samples. This limitation is an important issue, as (i) after each policy update, we must throw away the samples collected so-far and (ii) we cannot use a human expert to collect samples (iii) the policy π_θ does, in general, not guarantee any safe interaction with the environment.

Off-policy. Contrarily, an off-policy estimation of the gradient allows, in principle, any data usage and increasing sample efficiency In episodic reinforcement learning, one can use importance sampling to provide an off-policy estimate of the gradient,

$$\nabla_\omega \mathbb{E}_{\theta \sim p(\theta|\omega)} [R(\theta)] = \nabla_\omega \int R(\theta) p(\theta|\omega) \, d\theta \approx N^{-1} \sum_{i=1}^{N} \frac{p(\theta_i|\omega)}{p(\theta_i|\omega')} R(\theta_i) \nabla_\omega \log p(\theta|\omega) \qquad \text{with } \theta_i \sim p(\theta_i|\omega')$$

$$(2.17)$$

where $p(\theta|\omega')$ is the policy used to collect the data. In step based reinforcement learning, one can use a similar correction

$$\nabla_\theta J(\theta) \approx \sum_{t=0}^{T-1} \gamma^t \left(\prod_{i=0}^{t} \frac{\pi(a_i|s_i)}{q(a_i|s_i)} \right) Q^\pi(s_t, a_t) \nabla_\theta \log \pi(a_t|s_t),$$

$$(2.18)$$

where $q(a_t|s_t)$ is the policy used to collect the data. In Equations 2.17 and 2.18, importance sampling correction causes a high variance of the estimates, and, furthermore, hinders (i) the usage of deterministic policy and (ii) external sources of the dataset (i.e., collected from an unknown policy). To reduce the variance in episodic case, one can use self-normalized importance sampling (Owen, 2013) or baseline subtraction (Jie & Abbeel, 2010). However, in the step based scenario, the variance grows multiplicatively in the number of steps, making the estimation harder. In **online** reinforcement learning (where the agent is allowed to interact with the environment), the usage of off-policy estimates allows sample reuse enhancing sample efficiency. In **offline**

---

[3]The discount γ^t is often omitted in the current deep-learning literature, as noted by Nota & Thomas (2019).

reinforcement learning, instead, the system is not allowed to interact with the environment, entirely relying on a dataset provided by an external source (in our case a human expert). This technique is safer for the robot, but hinders the direct usage of importance sampling, as the human behavior is not known in analytic form. The difference between online, offline, on-policy and off-policy is summarized in Figure 2.5 To circumvent this handicap, in Chapter 5 we propose a novel off-policy gradient estimate based on a non-parametric technique. A richer analysis of state-of-the-art techniques can be found in Chapter 5 and in (Levine et al., 2020).

2.2. Experimental Setup

To support our theoretical analysis, we relied on numerical simulations and robotic experiments. In this section, we detail the technical settings used across this book.

2.2.1. Software Simulations

All the numerical simulations and the algorithm implementations contained in this book are entirely written in Python (Oliphant, 2007). To support numerical computations, we used Numpy and Scipy (Bressert, 2012). To use standard machine learning tools, such as k-means to initialize a Gaussian mixture, we used Scikit-Learn (Pedregosa et al., 2011). Automatic differentiation i s p erformed w ith P yTorch, and with few exceptions with TensorFlow (Paszke et al., 2017; Abadi et al., 2016). To provide task simulations, we used OpenAIGym and RLBench (which relies on CoppeliaSim) (James et al., 2020). For robotic communications, we used RosPy. The plots contained in this book are generated using Matplotlib and TikzPlotLib (Hunter, 2007). We provide a well-documented code for Chapter 4, Chapter 5, Chapter 6 at address `https://github.com/SamuelePolimi/LAMPO`, `https://github.com/jacarvalho/nopg`, and `https://github.com/SamuelePolimi/UpperboundNWBias` respectively.

2.2.2. Robotic Setup

To test our algorithms, we used a real cart-pole and a Lightweight KUKA arm with 7 DoF, as depicted in Figure 2.6.

Cart-Pole. The real Quanser cart-pole, consists of an actuated cart moving on a fixed track. The cart has attached a pole, which is free to move. By actuating the cart, one can swing the pole and eventually stabilize it in an upwards vertical position.

Robotic Manipulator: Our Lightweight KUKA arm comprises seven revolute joints, which allows a wide range of movements. The arm is equipped with an anthropomorphic DLR hand. The software controlling the robot offers the possibility to set the arm in gravity compensation mode, which allows a human demonstrator to guide the arm and record the desired movement. An optical tracking system, composed of six cameras, can track with millimetric precision the position of some markers in the scene. We also used a DYMO scale that interfaces with the computer in our pouring experiments. The scale has a precision of 2 grams.

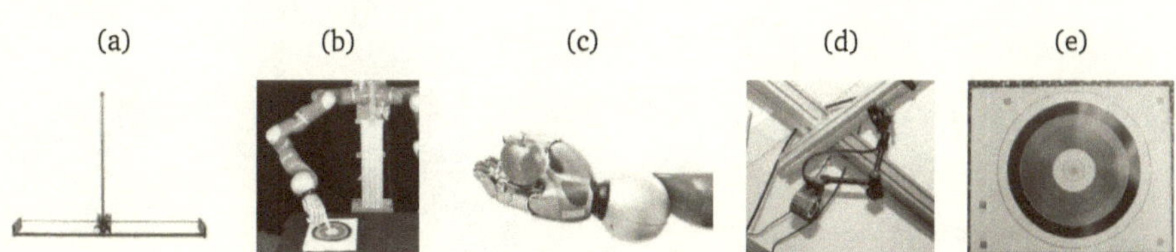

Figure 2.6.: (a) A real carp-pole platform. (b) Our robotic manipulator. (c) A zoom on the DLR hand. (e) A camera for the tracking system. (e) An object used in the experiments, equipped with markers recognizable by the tracking system.

2.3. Contributions

The goal of this book is to propose and analyze methods that facilitate automated learning for real robotics. To this end, as we analyzed, we must provide (i) safe and (ii) sample-efficient reinforcement learning.

In this book, we enhance safety in different ways. A first contribution is to analyze an efficient dimensionality reduction of movement parameters and to combine it to imitation learning. The key idea is that the projection to a latent space during the imitation phase enables the policy improvement in the latent space, and therefore an exploration in a region of safe policies, as we have seen in the Section 2.1.1. As a side benefit, reinforcement learning will also enjoy an enhancement in terms of sample efficiency, as the exploration is conducted in a smaller space.

Furthermore, in Chapter 5, we propose an off-line reinforcement technique that learns from a fixed dataset. In this way, the reinforcement learning system entirely relies on safe interactions provided by an expert. Furthermore, while state-of-the-art techniques require both a known stochastic behavioral policy and stochastic optimization policy, our method works also with deterministic optimization policy and completely undetermined behavioral policy.

The off-policy techniques introduced in Chapter 4 and 5 also help deliver a better sample efficiency, as the core idea of off-policy reinforcement learning is to reuse samples of past experiences to enhance learning further.

In the spirit of safety, we investigated an upper bound on the kernel regression bias used in Chapter 6. We believe that accompanying our findings with hard guarantees of the estimators' properties is crucial to obtain safe artificial intelligence.

Sample efficiency can also be seen as a matter of exploration/exploitation dilemma. An agent which explores not enough will probably converge to an inferior solution, while an agent that explores too much will produce unnecessary interaction with the system.

In Chapter 7, we provide a far-sighted exploration, designed appositely for tasks with a long horizon. As we will see, the main idea is to use an ensemble of estimators to estimate uncertainty and propagate optimistic estimations through the bellman recursion.

As a result of our investigations, we learned some user-defined tasks with a few hundred interactions with the robot. The method presented in Chapter 5, although not thoroughly tested on a robotic-system, was able to learn a simulated task, vastly improving on two sub-optimal demonstrations and without any further interaction.

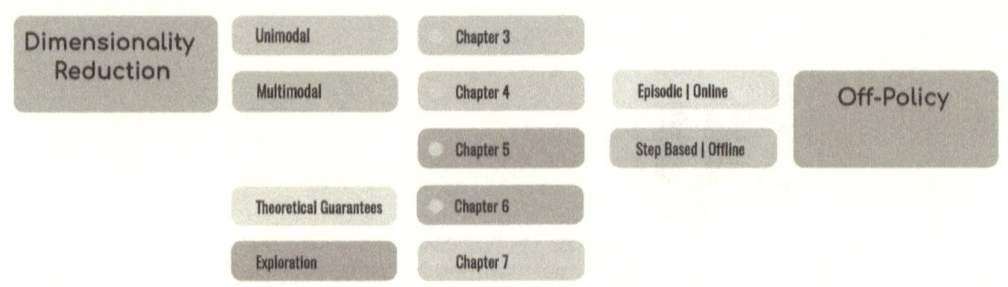

Figure 2.7.: The main contribution of this book relies on the usage of dimensionality reduction techniques to ensure a safer and efficient exploration while off-policy reinforcement learning ensures efficient sample reuse and allows to directly optimize on safely-collected data (Chapters 3, 4, 5). Chapter 6 provides theoretical guarantees on the core-method used in Chapter 5. In Chapter 7 we propose a novel exploration method which allows the usage of simple (sparse) reward functions and therefore require less human-knowledge.

2.3.1. Dimensionality Reduction of Movement Primitives in Parameter Space

Models proposed for learning robotic movements usually need a high number of parameters. The high dimensionality of the parameter space hinders an efficient subsequent optimization with reinforcement learning. In the literature, it is common to apply a dimensionality reduction technique in the robot's configuration space, which corresponds to the identification and subsequent compression of coupled joints. This model can be useful in a complex, highly redundant cinematic structure, such as the human body, while it loses its efficacy in simple robotic manipulators. In most manipulation tasks, while there is a low correlation in the joint space, we usually observe a strong correlation between movements. For this reason, we propose the application of principal component analysis (PCA) and its mixture version (MPPCA) in the parameter space, and we compare it with the dimensionality reduction technique, finding surprisingly that both complex structures (like the human body) and simple robotic manipulator, the dimensionality reduction works more efficiently in parameter space, allowing to reduce the dimension of the parameters of order of magnitudes.

2.3.2. Off-Policy Optimization for Robotic Manipulation Skills

In many robotic applications, the robot has to adapt its behavior to an observable context. The context can consist of some object's position in the scene or some goal dynamically defined by the user. The Probabilistic Movement Primitives (ProMPs) framework offers a tool to condition the movement given some context variable. In this work, we want to introduce a convenient dimensionality reduction in the parameter space; however, the proposed PCA in parameter space detailed in Section 2.1 is not effective in the presence of multimodalities. For this reason, we developed a dimensionality reduction technique in the parameter space based on the Mixture of Principal Component Analyzers (MPPCA) (Tipping & Bishop, 1999a). This technique's usage allows the identification of the different modalities contained in the dataset and still enables statistical conditioning in closed form. We empirically prove our imitation learning's effectiveness by learning with only 200 demonstrations a robotic task with 92-dimension of the context and over 120 parameters of the original movements. However, as discussed, the dimensionality reduction in imitation learning mainly serves to empower reinforcement learning. We propose a novel off-policy reinforcement learning policy optimization in the latent space of the MPPCA model based on self-normalized importance sampling. Furthermore, inspired by trust-region policy search methods like REPS and TRPO (Peters et al., 2010; Schulman et al., 2015), we ensure safe policy optimization using Kullback-Leibler (KL) constraints. We demonstrate the efficacy and efficiency of this novel off-policy technique in the movement's latent space by applying it to simulated and real robotic tasks. In all cases, our algorithm outperforms state-of-the-art techniques.

2.3.3. Offline Reinforcement Learning with a Nonparametric Off-Policy Policy Gradient

In Chapter 4, we successfully introduced an episodic off-policy estimation based on self-importance sampling (Owen, 2013). However, in step-based reinforcement learning, importance sampling suffers from a prohibitive high variance since it grows multiplicatively in the number of steps (Liu et al., 2018, 2019). To mitigate this issue, semi-gradient approaches omit a term in the expansion of the policy-gradient, still guaranteeing convergence in finite action-state Markov Decision Process with a tabular representation of the policy (Degris et al., 2012a). The theory of semi-gradient approaches vastly influenced the upcoming research, and several papers adopted this technique showing high sample efficiency compared to more classic on-policy algorithms (Silver et al., 2014; Lillicrap et al., 2016; Haarnoja et al., 2018). However, some further investigations, shown the inapplicability of semi-gradient approaches to strongly off-policy data, such as in off-line reinforcement learning, where the dataset is fixed (Imani et al., 2018; Fujimoto et al., 2019). To mitigate the high bias of semi-gradient approaches and the high variance of importance-sampling correction, we introduce a technique based on non-parametric techniques. We develop a non-parametric Bellman equation that has a closed-form solution. The solution is analytically differentiable, and therefore allows us to estimate the full-gradient. Our approach efficiently delivers a high-quality estimation of the gradient from a few samples (as shown in the empirical Section) and overcomes some limitations of importance-sampling approaches. Semi-gradient correction requires the perfect knowledge of the behavioral policy and stochastic policies. In contrast, our method allows any behavioral policy and any (stochastic or deterministic) differentiable policy in the optimization process. We accompany our method's derivation with both theoretical guarantees and a detailed empirical analysis of the bias and variance of our proposed estimator. Our method outperforms state-of-the-art off-policy algorithms in terms of sample-efficiency and remarkably solves simulated mountain-car tasks by learning from two sub-optimal demonstrations.

2.3.4. An Upper Bound of the Bias of Nadaraya-Watson Kernel Regression

Nadaraya-Watson kernel regression, (Nadaraya, 1964; Watson, 1964) is at the core of the reinforcement learning technique developed in Chapter 4. To develop the theoretical analysis of the value function's bias, we needed to develop an upper bound on the Nadaraya-Watson kernel regression. This upper-bound is one of the contributions of our book and is presented in Chapter 6. At the best of our knowledge, Nadaraya-Watson kernel regression bias has not been upper-bounded, except for very restrictive assumptions like evenly distributed data. Our bound is open to a broad class of design accomodating a weak-log-Lipschitz property, a broad class of weak-Lipschitz regression functions, and a wide class of kernel functions. We provide a detailed theoretical and empirical analysis of our bound, studying its behavior in limiting cases, comparing it through numerical evaluations to the actual bias, and Rosenblatt's asymptotic analysis (Rosenblatt, 1969). Our analysis shows that our bound is correct and tight and scales similarly to Rosenblatt's analysis, but being more robust at the presence of high second derivatives.

2.3.5. Exploration via an Optimistic Bellman Equation

An open problem in reinforcement learning is how to balance exploration and exploitation. While there is a vast amount of literature on the simplified problem of multi-armed bandits, where each action execution does not change the future context/state's distribution, the problem becomes less evident in classic MDPs. Fully exploring the state-action spaces in high-dimensional sparse-reward tasks is particularly challenging. If we consider the main deep-learning techniques applied to Atari games, we notice that they perform poorly in those games where exploration is required while successfully solving most other games (Mnih et al., 2015). The whole concept of exploration can be seen as the design of an additive term added to the Q-function proportional to the epistemic uncertainty. Most of the exploration techniques use a myopic additive bonus, reflecting the current uncertainty in a precise state-action pair.

We propose, instead, a novel technique based on an optimistic Bellman equation. The main idea is to provide an optimistic bonus under uncertainty and propagate it in the time horizon. This way, we can provide a bonus for action that will drive the agent to future state-action pairs with high uncertainty. We offer a detailed theoretical analysis of the algorithm, proving convergence, and showing how our technique can be related to state-count approaches by performing no direct estimates. We test our proposed method on a set of classic benchmarks, showing our method's efficacy compared to state-of-the-art techniques.

2.4. book Outline

Chapter 3 introduces an analysis of dimensionality reduction in the robot's configuration space compared to the movement's parameter space. The empirical analysis shows that the latter is more convenient, as it requires fewer dimension to reach high accuracy.

Chapter 4 introduces a novel algorithm called LAMPO (LAtent Movement Policy Optimization) that consists of applying MPPCA to the movement's parameter space, and in a novel, off-policy reinforcement technique applied in the movement's latent space. The method is tested in simulated and real robotics and outperforms the baselines in sample efficiency, solving high dimensional tasks with a few hundred samples.

Chapter 5 introduces NOPG, a novel off-policy learning algorithm that learns from a fixed dataset of human demonstrations. The guiding principle of the algorithm is the estimation of the gradient from a closed-form solution of a non-parametric Bellman equation, avoiding, therefore, the high bias of semi-gradient approaches and the high-variance of

Chapter 6 elaborates the bound on the bias mentioned in Chapter 5 by enriching it, analyzing it, and comparing it to state-of-the-art bias analysis. The analysis results show that our bound is correct, tight, and applies to a wide range of different conditions.

Chapter 7 presents a novel exploration algorithm that uses a new optimistic Bellman equation derived with an information-theoretic approach, which propagates an exploration bonus proportional to the epistemic uncertainty. It follows a theoretical guarantee of convergence within an analysis of the property of our exploration bonus. Eventually, the empirical analysis shows the superiority of our approaches when compared to less far-sighted exploration.

Chapter 8 provides a unified overview of our research's theoretical findings, details the remaining questions, and outlines the future investigation's direction.

3. Dimensionality Reduction of Movement Primitives in Parameter Space

Movement primitives are an important policy class for real-world robotics. However, the high dimensionality of their parametrization makes the policy optimization expensive both in terms of samples and computation. Enabling an efficient representation of movement primitives facilitates the application of machine learning techniques such as reinforcement on robotics. Motions, especially in highly redundant kinematic structures, exhibit high correlation in the configuration space. For these reasons, prior work has mainly focused on the application of dimensionality reduction techniques in the configuration space. In this paper, we investigate the application of dimensionality reduction in the parameter space, identifying *principal movements*. The resulting approach is enriched with a probabilistic treatment of the parameters, inheriting all the properties of the Probabilistic Movement Primitives. We further improve our method by introducing a *mixture of probabilistic principal movements* framework, which enables to deal with non-linearities in the data and to provide a probabilistic framework at once. We test the proposed technique on a database of complex human movements, on a set of demonstrations of diverse simulated robotic manipulation tasks, and on a real robotic task. The empirical analysis shows that the dimensionality reduction in parameter space is more effective than in configuration space, as it enables the representation of the movements with a significant reduction of parameters.

3.1. Prologue

Robot learning is a promising approach to enable more intelligent robotics, which can easily adapt to the user's desires. In recent years, the field of reinforcement learning (RL) has experienced an enormous advance in solving simulated tasks such as board- or video-games (Mnih et al., 2015; Lillicrap et al., 2016; Schulman et al., 2017), in contrast to little improvements in robotics. The major challenges in the direct application of RL to real robotics, are mainly the limited availability of samples and the fragility of the system, which disallow the application of unsafe policies. These two disadvantages become even more evident when we consider that usual robotic tasks such as industrial manipulation, are defined in a high dimensional state and action space. A usual approach in the application of RL to robotics, is to initialize the policy via imitation learning (Schaal, 1999; Billard & Siegwart, 2004; Argall et al., 2009; Rana et al., 2018). In the past, there has been some effort in providing a safe representation of the policy for robotics, mainly by the means of Movement Primitives (MPs) (D'Avella et al., 2003; Schaal et al., 2005; Khansari-Zadeh & Billard, 2011; Paraschos et al., 2013). The general framework of MPs has been extensvely studied and employed in a large variety of settings (Amor et al., 2014; Maeda et al., 2014; Koert et al., 2016; Maeda et al., 2017; Stark et al., 2019). MPs have been shown to be effective when used for the direct application of RL to robotics (Peters & Schaal, 2008; Kober & Peters, 2009; Mülling et al., 2013). The drawback of the general framework of MPs, is the usual high number of parameters. In fact, in the robot's configuration space, the number of parameters is equal to the

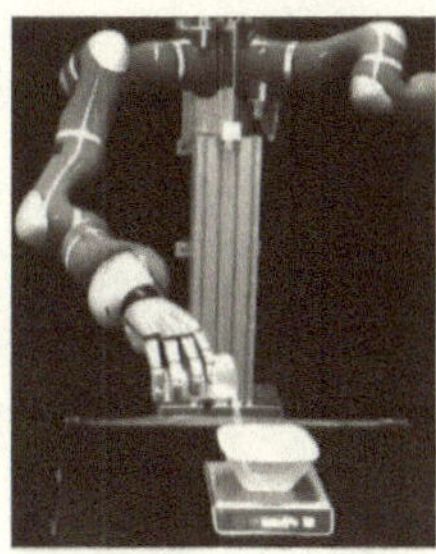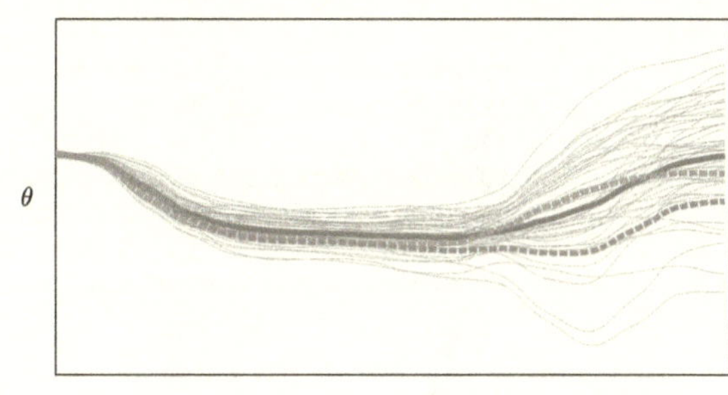

Figure 3.1.: Our binamual platform performing a pouring task using Pro-Primos (left). The generation of trajectories via PriMos (right): the blue line corresponds to the mean movement, while the dashed lines to the principal movements. By only linearly combining two principal movements, it is possible to generate a wide variety of movements (gray lines).

number of basis functions (typically greater than 10), times the number of degree-of-freedom (DoF). The number of parameters grows even quadratically w.r.t. the number of basis function and DoF when we want to represent the full covariance matrix in the case of Probabilistic Movement Primitives (ProMPs) (Paraschos et al., 2013, 2018). In recent years, many authors have proposed techniques to solve the problem by using a latent representation of the robot's kinematic structure (Colomé & Torras, 2014; Colomé et al., 2014; Chen et al., 2015, 2016). Especially for complex systems, such as humanoids, many DoF are redundant, and a compressed representation of the joint-space[1] results in a lighter parametrization of the MP. However, this approach is less effective in the case of a robotic arm with fewer DoF Moreover, the parametrization of the MPs would be still linear w.r.t. the number of basis functions.

In this chapter, we build on the general framework of ProMPs and we propose to apply the dimensionality reduction in parameter space. In tasks where there is a high correlation between the movements, it makes more sense to seek a compressed representation of the movement instead of the configuration space. To this end, we propose an approach where the movements can be seen as a linear combination of *principal movements* (Figure 3.1). We enrich our framework with a probabilistic treatment of the parameters, so that our approach inherits all the properties of ProMPs. Furthermore, we also propose a mixture of probabilistic principal movements (PMMPs) that overcomes the limitations of PriMos and Pro-PriMos. We analyze the benefit of dimensionality reduction in the parameter space w.r.t. in joint space, on a challenging human motion reconstruction, on a set of simulated manipulation robotic tasks and on a real robot performing a pouring task (Figure 3.1). Our findings show that our proposed approaches achieve satisfying accuracy with a significant reduction of parameters.

3.1.1. Problem Statement

Machine learning, should allow the robotic agent to interact with the real world in a non-predetermined way. However, the possible generated movement should be smooth, and possibly constrained to be safe (not colliding with other objects, without excessive speed or accelerations, etc). These issues require a whole field of research to be solved. MPs are useful in this context. The idea behind MPs, is to restrict the class

---

[1] In this chapter, we use joint space and configuration space interchangeably.

of all possible robotic movements to a specific parameterized class usually by linearly combining a set of parameters with a set of features. For a particular class of features (e.g., radial basis function), the movements are guaranteed to be smooth. However, sometimes restricting the space of robotic movement to be smooth is not enough, and we want also the movements to be distributed similarly to some demonstration provided by a human expert. The ProMPs provides a probabilistic treatment of the movement's parameters (Paraschos et al., 2013, 2018). Both frameworks are useful and can be coupled with RL. MPs and their probabilistic version (ProMPs) require a high parametrization, which makes them of difficult use in reinforcement learning.

3.1.2. Movement primitives

Given a robot with d degrees of freedom (dof), for which we can observe a trajectory $\tau = \{y_1, ..., y_T\}$ for an episode of duration T, where $y_t \in \mathbb{R}^d$ is the robot's joint configuration at time t (or any other quantity, i.e. Cartesian positions of the robot's end-effector). An MP is a parameterized model with

$$y_t = \Psi_t \omega + \varepsilon_y \tag{3.1}$$

where $\Psi_t \in \mathbb{R}^{d \times (d \cdot n)}$ with $\Psi_t = \Phi_t \otimes I_d$, i.e. the Kronecker product of the d-dimensional identity matrix and the feature vector $\Phi_t = [\phi_{1t}, ..., \phi_{nt}]$. $\omega \in \mathbb{R}^{d \cdot n}$ are weights on the feature vector, and $\Phi_t \in \mathbb{R}^d$, and ϕ_{it} is a shortcut for a normalized radial basis function, i.e., $\phi_{it} = b_i(t)/\sum_{j=1}^n b_j(t)$, while $\varepsilon_y \sim \mathcal{N}(0, \sigma^2 I)$, is a zero-mean Gaussian noise, representing noisy observations. The problem of finding the best parameters for representing the MPs in (3.1) can be seen as a *maximum-likelihood estimation* (MLE) problem

$$\max_\omega \mathcal{L}(\omega) = \max_\omega \prod_t \mathcal{N}(y_t \mid \Psi_t \omega, \Sigma_y). \tag{3.2}$$

which can be solved with linear regression.

3.1.3. Probabilistic Movement Primitives

Probabilistic Movement Primitives (ProMPs) is a flexible probabilistic framework for representing motor policies Paraschos et al. (2018), having desirable MPs properties like co-activation, conditioning, etc. Aparting from the formalization of the general regression problem of (3.1), ProMPs encode a the distribution of the trajectories τ, using as feature encoders basis-function representations. For *time modulation*, we use a *phase*-vector z where $z_i = (t_i - t_1)/(t_T - t_1)$, t_i being the ith timestep. From the general MPs formalization in (3.1), we consider n normalized radial-basis functions Φ_t, usually, ordered to evenly cover the phase-space, and centered between $[-2h, 1 + 2h]$ where h is the bandwidth. At time t the features are described by a n-dimensional column vector Φ_t where $\Phi_{t_i} = \Phi_i(z_t)$. In ProMPs, the weights ω are treated as random variables with distribution $p(\omega) = \mathcal{N}(\omega|\mu_\omega, \Sigma_\omega)$ allowing a probabilistic treatment of the MPs. In this case, we want to find the best distribution over parameters which encodes a distribution of trajectories, i.e.,

$$p(\tau \mid \mu_\omega, \Sigma_\omega) = \int p(\tau \mid \omega) \mathcal{N}(\omega \mid \mu_\omega, \Sigma_\omega)\, d\omega,$$

$$= \int \mathcal{N}(y_{1:T}|\Psi_{1:T}\omega, \Sigma_y)\mathcal{N}(\omega|\mu_\omega, \Sigma_\omega)\, d\omega,$$

$$= \mathcal{N}(y_{1:T}|\Psi_{1:T}\omega, \Psi_{1:T}\Sigma_\omega \Psi_{1:T}^\mathsf{T} + \Sigma_y). \tag{3.3}$$

Imitation Learning with ProMPs

For a finite set of stroke-movements $\{\tau_i\}_{i=1}^{m}$, it is possible to use the parameters ω_i estimated for each trajectory τ_i in order to estimate μ_ω and Σ_ω, i.e.,

$$\mu_\omega = \frac{1}{m} \sum_{i=1}^{m} \omega_i, \quad \Sigma_\omega = \frac{1}{m} \sum_{i=1}^{m} (\omega_i - \mu_\omega)(\omega_i - \mu_\omega)^\mathsf{T}. \tag{3.4}$$

across all trajectories, which can be learned from data by maximum likelihood using the Expectation Maximization (EM) algorithm. The advantage of this probabilistic approach is the possibility to learn from incomplete data, i.e. due to observations losses during the demonstrations. However, the training of ProMPs presents from a severe disadvantage. As the model has a lot of parameters due to the high-dimensional covariance matrix, ProMPs suffer from overfitting if we have little training data and noisy trajectories.

Contextual Conditioning of ProMPs

An important property of ProMPs is their modulation via probabilistic conditioning. In particular, adaptation to different contexts (high-level parameters that encode an aspect of the task, e.g. the target's poisition in a reaching task) is an important requisite for MPs, so that they are able to generalize to different settings of the demonstrated task. In particular, given a context c, modelled as a Gaussian distribution, we aim to learn the distribution of parameters ω conditioned on c, i.e., $p(\omega|c)$, which is a Gaussian distribution with mean Therefore, suppose that the movement parameters and the context are jointly Gaussian distributed. Suppose $\mu_c = \mathbb{E}[c]$, $\Sigma_{\omega,c}$ being the covariance matric between ω and c, and Σ_c being the covariance of the context. We can derive that $p(\omega|c) = \mathcal{N}(\omega|\mu_{\omega|c}, \Sigma_{\omega|c})$ with

$$\mu_{\omega|c} = \mu_\omega + \Sigma_{\omega,c}\Sigma_{c,c}^{-1}(\omega - \mu_\omega),$$
$$\Sigma_{\omega|c} = \Sigma_\omega - \Sigma_{\omega,c}\Sigma_c^{-1}\Sigma_{\omega,c}^\mathsf{T}.$$

The contextualized conditional distribution $p(\omega|c)$ is used to predict the robot trajectories w.r.t. to the considered context of the executed task.

3.1.4. Dimensionality Reduction

Due to the high-dimensionality of robotic manipulation tasks, in particular for learning from demonstrations, it has been a common practice to reduce the space of the demonstrated data. As the parameters needed to encode the MPs might be even more that the dofs of the task, we seek to project the demostated data points to a subspace, a *latent* space, that can encode the important correlation in the data. For ProMPs the extraction of the latent space can be performed through dimensionality reduction (DR) in the joint space of the demonstrated trajecories, or in the parameter space of the encoded trajectories.

Dimensionality Reduction in Joint Space. In principle the problem of DR, lies in finding the mapping $y_t' = f_J(y_t)$, with $f_J : \mathbb{R}^d \to \mathbb{R}^{d_2}$ with $d_2 \leq d$. Hence, (3.1) is mapped to the reduced manifold of f:

$$y_t' = \Psi'(t)\omega' + \epsilon' \tag{3.5}$$

where $\omega' \in \mathbb{R}^{d_2 \cdot n}$ the parameters of the *latent* movements, the reduced trajectories $y' \in \mathbb{R}^{d_2 \times (d_2 \cdot n)}$ and the reduced features $\Psi'(t) = \Phi(t) \otimes I_{d_2}$. The movement can be reconstructed using $\tilde{f}_J$, i.e., $y_t = \tilde{f}_J(y_t')$.

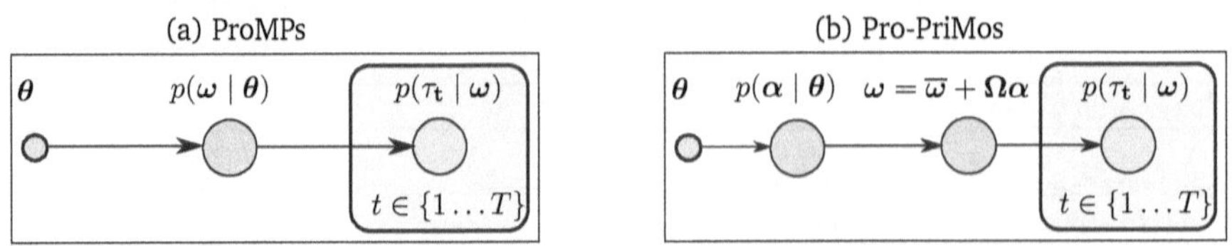

Figure 3.2.: Graphical model of the ProMPs framework (a) compared to Pro-PriMos (b). The main idea of Pro-PriMos is to generate a distribution of linear combinations of the principal-movements Ω.

Dimensionality Reduction in Parameter Space. For mapping the ProMPs to a latent space, we would in principle map the space of the MP parameters to a reduced one, $\omega' = f_P(\omega)$, with $\omega' \in \mathbb{R}^{d_2}$ with $d_2 < d \cdot n$. If $\tilde{f}_P$ is the reverse mapping, we can reconstruct the trajectories as: $y_t = \Psi_t \tilde{f}_P(\omega') + \varepsilon_y$. This approach is more convenient for two main reasons: first, working in the parameter space we can control how many parameter to use, while in the joint space, we always obtain a number of parameters proportional to the number of RBFs, second, the parameters encodes higher correlations, as we will see in the empirical analysis.

3.1.5. Related Work

The issue of dimensionality reduction in the context of motor primitives has been extensively studied. To achieve dimensionality reduction for Dynamic Movement Primitives (DMPs) (Schaal et al., 2005), the *autoencoded dynamic movement primitives* model proposed in (Chen et al., 2015), uses to find a representation of the movements in a latent feature space, while in (Chen et al., 2016) the DMPs are embedded into the latent space of a time-dependent variational autoencoder. In (Colomé et al., 2014) a linear projection in the latent space of the configuration space is considered, as well as the adaption of the projection matrix in the RL context. Interestingly, they consider the possibility to address the dimensionality reduction in the parameter space, but they discard this option as the projection matrix is more difficult to adapt since it is of higher dimensions. We agree with this argument, however, when we assume that the projection matrix can be considered fixed this argument is not valid anymore. The dimensionality reduction of ProMPs has been addressed in (Colomé & Torras, 2018a), where the authors compare PCA versus an expectation-maximization approach in configuration space. Until this point, all the literature is focused in finding a mapping between configuration space and a latent space, and proposing the learning of the MPs in this lower representation. However, this approach results to be more efficient in complex kinematic structures, such as the human body, where the high number of joints facilitate the possibility of redundancies in the configuration space. Moreover, this reduction, does not affect the intrinsic high-dimensional nature of the MPs, which requires usually a high number of basis functions.

A way to overcome this problem is to focus in the parameter space of the movement primitives. The dimensionality reduction in this case exploits similarities between movements, and high correlations between parameters. To the best of our knowledge, the only work performing a reduction in parameter is (Rueckert et al., 2015). The proposed setup is however complex as it considers a fully hierarchical Bayesian setting, where the movements are encoded by a mixture of Gaussian models. They learn those parameters using variational inference. Furthermore, their approach does not address the question of whether the parameter space reduction is more convenient or not.

In our paper, we want to focus on the comparison between the dimensionality reduction in configuration and parameter space, arguing that the latter is more convenient. To this end, we propose the Principal Movements (PriMos) framework, which enables the selection of *principal movements* and the subsequent representation of the movements in this convenient space. We extend PriMos to incorporate a probabilistic treatment of the parameters (Pro-PriMos), in a similar way to the ProMPs framework. This approach only adds a linear transformation to the framework already developed by Parachos et al., as depicted in Figure 3.2, therefore it maintains all the properties exposed by the ProMPs, such as time modulation, movements co-activation or movement conditioning. We further enhance our model proposing a mixture of probabilistic principal movement to deal with non-linearities in the data.

3.2. The Principal Movements Framework

To reduce the dimensionality of MPs, we propose to apply dimensionality reduction in the parameter space. We introduce the Principal Movements (ProMPs), which are derived applying PCA to the parameter space of movement primitives. We extend this framework with a probabilistic treatment of the parameters, introducing Pro-PriMos in Section 3.2.2. We can appreciate the difference between ProMPs and Pro-PriMos in Figure 3.2. PriMos and Pro-PriMos exploit only linear dependencies between the data, strongly limiting their applicability to complex scenario. We introduce the Mixture of Probabilistic Principal Movements (MPPMs) in Section 3.2.3, which at once overcome the aforementioned limitation, and naturally introduce a probabilistic view. Eventually, we can see that PriMos can be seen as a special case of MPPMs.

3.2.1. Principal Movements

The MPs and ProMPs frameworks, usually require a large number of parameters for encoding the movements. In MPs, we need in fact $d \times n$ parameters (where d is the dimension of the considered movement and n is the number of radial basis functions). The choice of the number of radial-basis functions usually depends by both the speed and the complexity of the movements, but most of the time it is greater than ten. For a 7 d.o.f. we, therefore, need at least 70 parameters to encode the MPs. This makes the application of RL challenging for a 7 d.o.f. robot arm. We will assume, from now on, that an oracle (i.e., a human expert or a dimensionality reduction method) will give us the parameters of a *mean-movement* $\overline{\omega}$, and a matrix of *principal-movements* $\Omega^\mathsf{T} = [\omega_1^p, \ldots, \omega_{n_c}^p]$. We therefore want to encode a given trajectory τ as a linear combination of the principal movements, i.e.,

$$\tau_t = \Psi_t \overline{\omega} + \Psi_t \Omega \alpha + \epsilon_t. \tag{3.6}$$

We consider $\alpha \in \mathbb{R}^{n_c}$ the new parameter vector. Note that n_c is generally independent of the number of joints d and of the number of radial basis function n, but instead the choice of n_c is connected to the *complexity* of the movement-space that we aim to represent. The MLE problem

$$\max_\alpha \prod_{t=1}^{T} \mathcal{N}(\tau_t \mid \Psi_t \overline{\omega} + \Psi_t \Omega \alpha, \sigma_t), \tag{3.7}$$

induced by (3.6) has a Ridge regression solution

$$\alpha = \left(\Omega^\mathsf{T} \Psi^\mathsf{T} \Psi \Omega + \lambda \mathbb{I} \right)^{-1} \Omega^\mathsf{T} \Psi^\mathsf{T} \left(\tau - \Psi \overline{\omega} \right). \tag{3.8}$$

3.2.2. Probabilistic Principal Movements

A probabilistic treatments of the MPs, is often convenient, as it enables the application of statistical tools (Paraschos et al., 2013, 2018). To enrich our approach with a probabilistic treatment, we assume our parameter

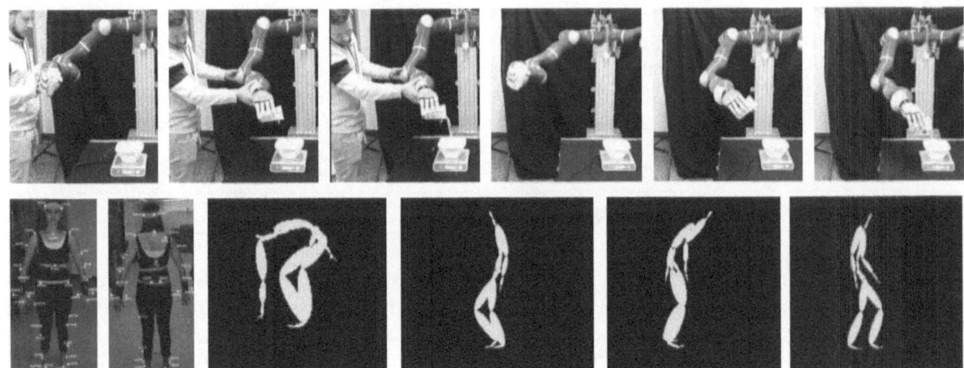

Figure 3.3.: Top: From left to right, the demonstration in kinestatic teaching mode of the pouring task, and the subsequent reconstruction of the movement with PriMos. Bottom: Markers placements on the human body, and illustrative example of the skeleton reconstruction for a "walk and pick-up" task of subject #143 of the MoCap database `http://mocap.cs.cmu.edu`.

vector α to be multivariate-Gaussian distributed, i.e., $\alpha \sim \mathcal{N}(\cdot \mid \boldsymbol{\mu}_\alpha; \boldsymbol{\Sigma}_\alpha)$. Very similarly to the classic ProMPs approach, we have that

$$p(\boldsymbol{\tau}_t \mid \boldsymbol{\mu}_\alpha, \boldsymbol{\Sigma}_\alpha) = \int \mathcal{N}(\boldsymbol{\tau}_t \mid \boldsymbol{\Psi}_t(\boldsymbol{\Omega}\alpha + \overline{\varpi}), \boldsymbol{\Sigma}_\tau)\mathcal{N}(\alpha \mid \boldsymbol{\mu}_\alpha, \boldsymbol{\Sigma}_\alpha)\,\mathrm{d}\alpha$$

$$=\mathcal{N}(\boldsymbol{\tau}_t \mid \boldsymbol{\Psi}_t(\boldsymbol{\Omega}\boldsymbol{\mu}_\alpha + \overline{\varpi}), \boldsymbol{\Psi}_t\boldsymbol{\Omega}\boldsymbol{\Sigma}_\alpha\boldsymbol{\Omega}^\mathsf{T}\boldsymbol{\Psi}_t^\mathsf{T} + \boldsymbol{\Sigma}_\tau). \tag{3.9}$$

The mean $\boldsymbol{\mu}_\alpha$ and the covariance $\boldsymbol{\Sigma}_\alpha$ can be estimated as similarly done to ProMPs, (Equation 3.4)

$$\boldsymbol{\mu}_\alpha = \frac{1}{m}\sum_{i=1}^{m}\alpha_i, \quad \boldsymbol{\Sigma}_\alpha = \frac{1}{m}\sum_{i=1}^{m}(\alpha_i - \boldsymbol{\mu}_\alpha)(\alpha_i - \boldsymbol{\mu}_\alpha)^\mathsf{T}, \tag{3.10}$$

where the parameters α_i correspond to the trajectories $\boldsymbol{\tau}_i$.

We assume an affine transformation between the full parameters ω and the reduced α, as shown in Figure 3.2, and α is assumed to be Gaussian distributed. Under these assuptions, ω is also Gaussian distributed, allowing for a mapping from Pro-PriMos to ProMPs,

$$\hat{\boldsymbol{\mu}}_\omega = \boldsymbol{\Omega}\boldsymbol{\mu}_\alpha + \overline{\varpi}, \qquad \hat{\boldsymbol{\Sigma}}_\omega = \boldsymbol{\Omega}\boldsymbol{\Sigma}_\alpha\boldsymbol{\Omega}^\mathsf{T}.$$

Hence, the proposed framework enjoys all the properties of the ProMPs, such as movements co-activation or movement conditioning.

3.2.3. Mixture of Probabilistic Principal Movements

The aforementioned model, relies on the assumption of Gaussian-distributed data, i.e., data with one mode and second order correlations between the variables. Furthermore, the probabilistic version is artificially built after the dimensionality reduction. This limitations of the Principal Movements can be overcome with the introduction of a Mixture of Probabilistic Principal Component Analyzers (MPPCA) (Tipping & Bishop, 1999a). The concept behind MPPCA is to use a probabilistic formulation of the PCA, and to extend it with a mixture model maintaining a full probabilistic description of the variables. This treatment, allows to use a generic expectation-maximization algorithm to infer the model's parameters. Furthermore, while PCA does only allow dimensionality reduction of the data, the probabilistic view grant us the access to probabilistic techniques, like statistical conditioning. This improvement is important as it allows to perform dimensionality reduction and imitation learning at once. In fact, one can include a *context* variable alongside the movement parameters, and subsequently, sample a movement conditioned on a specific observed context. In the following, we introduce the MPPCA in parameter space. We include the context variable $\mathbf{c} \in \mathbb{R}^{d_c}$ where d_c is the context's dimension, which can be used for contextual movement. The Mixture of Probabilistic Principal Movements (MPPMs) is described as

$$
\begin{aligned}
k &\sim Cat(k|\boldsymbol{\pi}), & (k \in \{1, \ldots K\}) \\
\mathbf{z} &\sim \mathcal{N}(\mathbf{z}|\mathbf{0}, \mathbf{I}), & (\mathbf{z} \in \mathbb{R}^{d_z}) \\
\epsilon_\omega, \epsilon_c &\sim \mathcal{N}(\epsilon_\omega, \epsilon_c|\mathbf{0}, \mathbf{I}), & \\
\omega &= \Omega_k \mathbf{z} + \overline{\omega}_k + \sigma_k^2 \epsilon_\omega, & (\omega \in \mathbb{R}^m) \\
\mathbf{c} &= \mathbf{C}_k \mathbf{z} + \overline{\mathbf{c}}_k + \sigma_k^2 \epsilon_c, & (\mathbf{c} \in \mathbb{R}^{d_c})
\end{aligned}
$$

where k is a categorical latent variable representing the selection of a particular linear transformation of a standard multivariate distribution $\mathbf{z}$. The definition of the isotropic noises $\epsilon_\omega, \epsilon_c$ allows to extract only the useful information, and act as a denoiser in the generative model.

Dimensionality Reduction. There are mainly two ways in which we can extract the optimal values of $\mathbf{z}$ and k given a parameter-context pair $(\omega, \mathbf{c})$: by maximum likelihood, i.e., $\max_{\mathbf{z},k} p(\omega, \mathbf{c}|\mathbf{z}, k)$ or by the posterior distribution, i.e., $\max_{\mathbf{z},k} p(\mathbf{z}, k|\omega, \mathbf{c})$. The latter approach is more natural in this probabilistic setting, furthermore, the optimization problem is particularly simple, as $\arg\max_{\mathbf{z}} p(\mathbf{z}|\omega, \mathbf{c}, k) p(k|\omega, \mathbf{c}) = \arg\max_{\mathbf{z}} p(\mathbf{z}|\omega, \mathbf{c}, k)$, and $p(\mathbf{z}|\omega, \mathbf{c}, k)$ is normally distributed. Let us introduce $\mathbf{A}_k = [\Omega_k, \mathbf{C}_k]^\mathsf{T}$, $\mathbf{a}_k = [\overline{\omega}_k, \overline{\mathbf{c}}_k]^\mathsf{T}$ and $\mathbf{x} = [\omega, \mathbf{c}]^\mathsf{T}$. The posterior distribution $p(\mathbf{z}|\mathbf{x}, k)$ has mean $\Sigma_k A_k^\mathsf{T}(\mathbf{x} - \mathbf{a}_k)$ with $\Sigma_k = (\mathbf{A}_k^\mathsf{T}\mathbf{A}_k + \sigma_k^2 \mathbf{I})^{-1}$. Hence, $\max_{\mathbf{z},k} p(\mathbf{z}, k|\omega, \mathbf{c})$ can be solved using the posterior mean $\mathbb{E}[\mathbf{z}|\mathbf{x}, k]$. The optimal reconstruction can be obtained by $\mathbf{x} = \mathbf{A}_k \mathbf{z} + \mathbf{a}_k$.

Movement Conditioning. We can use MPPM to infer the distribution $p(\omega|\mathbf{c})$. In fact, $p(\omega|\mathbf{c}_k, k)$ is normal distributed with

$$
\begin{aligned}
\mathbb{E}[\omega|\mathbf{c}] &= \overline{\omega}_k + \Omega_k \mathbf{C}_k^\mathsf{T}(\mathbf{C}_k \mathbf{C}_k^\mathsf{T} + \sigma_k^2 \mathbf{I})^{-1}(\mathbf{c} - \overline{\mathbf{c}}_k), \\
\mathrm{Cov}[\omega|\mathbf{c}] &= \Omega_k \Omega_k^\mathsf{T} - \Omega_k \mathbf{C}_k(\mathbf{C}_k \mathbf{C}_k^\mathsf{T} + \sigma_k^2 \mathbf{I})^{-1} \mathbf{C}_k \Omega_k^\mathsf{T} + \sigma_k^2 \mathbf{I},
\end{aligned}
\tag{3.11}
$$

hence, $p(\omega|\mathbf{c})$ results to be a mixture of Gaussian,

$$
p(\omega|\mathbf{c}) = \sum_{k=1}^{k} p(\omega|\mathbf{c}_k, k) p(k|\mathbf{c})
$$

with $p(k|\mathbf{c}) = p(\mathbf{c}|k)p(k) / \sum_i (p(\mathbf{c}|i)p(i))$ being a vector of responsabilities. Note that, the isotropic noise in $\mathrm{Cov}[\omega|\mathbf{c}]$ can be removed, so to generate a cleaner movements.

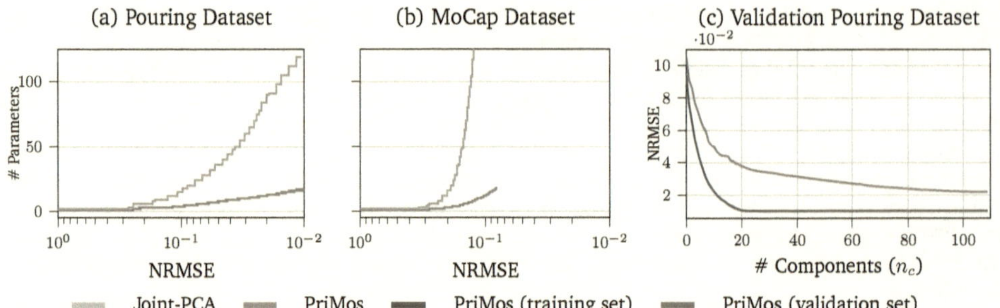

Figure 3.4.: The number of parameters (y-axis) needed to achieve a certain accuracy (x-axis) (a), (b). As can be seen PriMos achieve better results with less samples. The error has been normalized in order to highlight the difference in complexity between the two datasets. In plot (b), both PriMo and Joint-PCA fail in reaching a NRMSE lower than 0.08. This is due to the complexity of the movement in the dataset, and by the maximum number of radial basis functions (20). In plots (c) we observe the comparison between train (blue) and validation error (green) achieved by PriMos in the pouring datasets.

3.3. Empirical Analysis

Our main claim is that dimensionality reduction is more efficient when applied in the parametric space of Movement Primitive. To support this claim, we use three different dataset: a human's motion dataset (MoCap), six datasets of simulated robotic manipulation tasks using RLBench and eventually, we collected some data on a real robotic tasks, using our robotic manipulator.

The Human Motion Dataset. The MoCap database contains a wide range of human motions (such as running, walking, picking up boxes, and even dancing) using different subjects (Figure 3.3). Human motion is tracked using 41 markers with a Vicon optical tracking system. The data is preprocessed using Vicon Bodybuilder in order to reconstruct a schematic representation of the human body and the relative joint's angles including the system 3D system reference, for a total of 62 values. The human motion is known to be highly redundant (i.e. many configurations are highly correlated), and therefore this is in principle the best case for the dimensionality-reduction in the configuration-space. The presence of many different typologies of movements, makes the application of PriMos challenging since our algorithm relies on the correlation between movements. In our experiments, we use the 42 movements recorded for the subject #143. Furthermore, we use walk, a subset of the MoCap dataset containing different subjects in the act of walking.

RLBench Demonstrations. We use the RLBench framework to generate demonstration on 6 different task: close_drawer, pick_up_cup, reach_target, unplug_charger, water_plants and wipe_desk. The tasks are solved with a 6 DoF robotic arm equipped with a simple gripper. As we encode the gripper's opening with a scalar variable, the total dimension of the movement is 7.

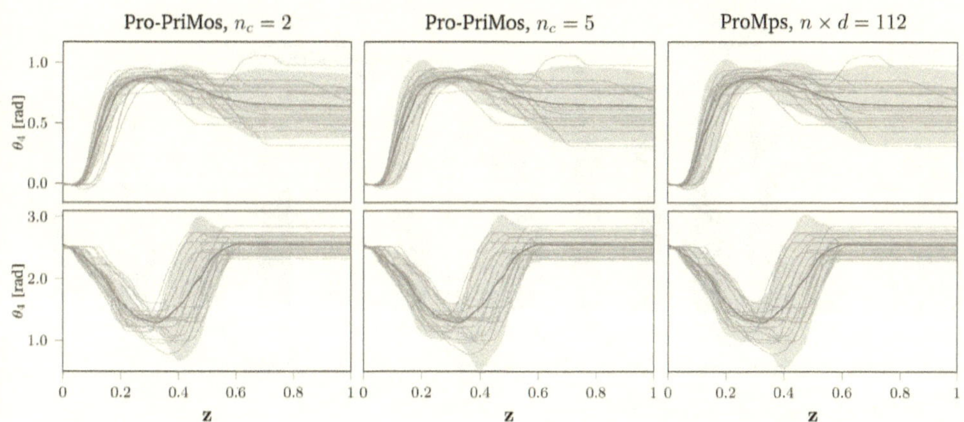

Figure 3.5.: Mean movement of the joints most involved in the movements (shoulder and elbow), accompanied by the standard deviation ($\pm 2\sigma$) for Pro-PriMos with 2 and 5 components, and with ProMPs with 112 parameters for the mean. Note that the covariance matrix is encoded with 4, 25 and 12544 values respectively.

The Pouring Task. In the pouring task, we use a KUKA light-weight robotic arm with 7 DoF accompanied with a DLR-hand as an end-effector to pour some "liquid" (which, for safety reasons, is replaced by granular sugar). We record some motions from a human demonstrator, pouring some sugar in a bowl. The motion is recorded setting the robotic arm in kinestatic teaching. The quantity of sugar contained in the bowl is recorded by a DYMO digital scale with a sensitivity of $2g$. We aim to reconstruct the movements and understand whether our method is able to pour a similar amount of sugar: this experiment gives us a qualitative understanding, beyond the numerical accuracy analysis, to investigate the effectiveness of our algorithm in a real robot-learning task.

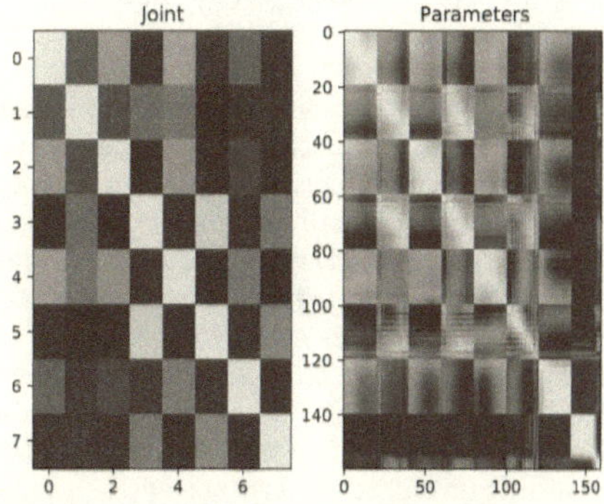

Figure 3.6.: An example of correlation between joints and between parameters. Higher correlations are denoted in yellow, and lower in dark blue. The correlation between parameters express both the correlation between joints and includes the time correlation.

3.3.1. Joint vs Parameter Correlation

To motivate the dimensionality reduction in parameter space, we perform a study on the correlation between joints and parameters on the RLBench set of tasks, and a subset of the MoCap dataset containing walk samples. Figure 3.6 depicts the correlation between joints and between parameters. While in joint the temporal information is lost, the parameter space captures both redundancies between joints and temporal dependencies. Figure 3.7 shows how the correlation results to be always higher in the parameter space. This higher correlation translates in higher possibility for

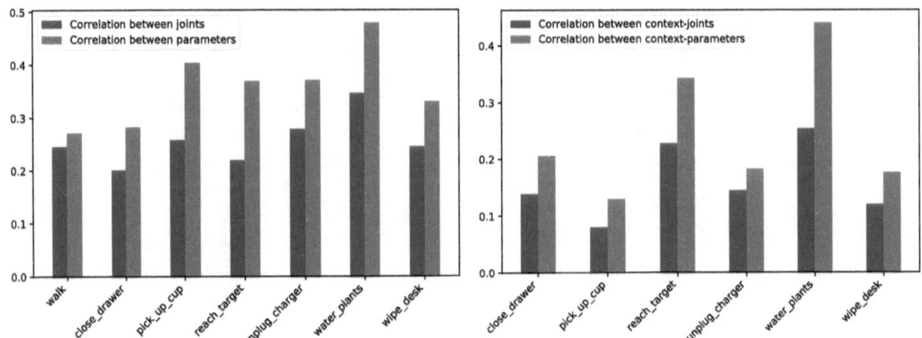

Figure 3.7.: A comparison between the correlation in joint and in parameter space. The correlation in the parameter space is higher. We observe a higher correlation also between context and parameters w.r.t. context and joints.

3.3.2. Accuracy of Movement Reconstruction

The higher correlation in the parameter space suggests that dimensionality reduction can potentially be more effective in parameter space w.r.t. the joint space. We test this hypothesis both with and without the context variable.

Without Context. In a first experiment, we want to compare PCA in joint space (Joint-PCA) and in parameter space (PriMos) on the reconstruction error. To this end, we consider a different number of parameters: in detail, when the dimensionality reduction is applied to the joint space, the resulting number of parameters can be obtained with different configurations of the dimensionality reduction technique and the number of features. Therefore, we evaluate all configurations that result in the same number of parameters, and we plot the lowest error achieved. In this way, we provide an optimistic (lower) bound for the dimensionality reduction in parameter space. We consider two datasets: the `Pouring Dataset` and the `MoCap`. We observe in Figure 3.9 that the dimensionality reduction results more effective in parameter space in both the datasets. Furthermore, no observed "over-fitting" effect shows that a simple model, like PCA, tends to generalize well. We compare PCA, PPCA, MPPCA, and a simple one-layer autoencoder (AE) in the joint and parameter space to further investigate different techniques. In this case, we keep the number of radial basis functions fixed at 20. The dataset is split into 80% train-set and 20% test-set. We evaluate the reconstruction on the test-set on the `RLBench` tasks. In Figure 3.8a one can observe that the MPPCA in the parameter space is very effective, reaching with few parameters one order of magnitude higher accuracy w.r.t. dimensionality reduction in joint space. Interestingly, when applied in the joint space, the same technique performs particularly bad, showing that the correlation is simple and can be extrapolated with simple linear DR. In Appendix, we report a more detailed analysis for each task.

With Context. In this experiment we take into consideration the contexts. In this case, in the joint space, we perform classic dimensionality reduction, and subsequently we fit a Gaussian distribution $p(\omega, c)$ over movement's parameters and contexts. In this way, we can infer $p(\omega|c)$. In the parameter space, we simply perform PPCA and MPPCA. The inclusion of the context, sensibly complicates the learning, as the majority of tasks have a high-dimensional context (around 90 dimensions). In this case, there is not a sensible difference in

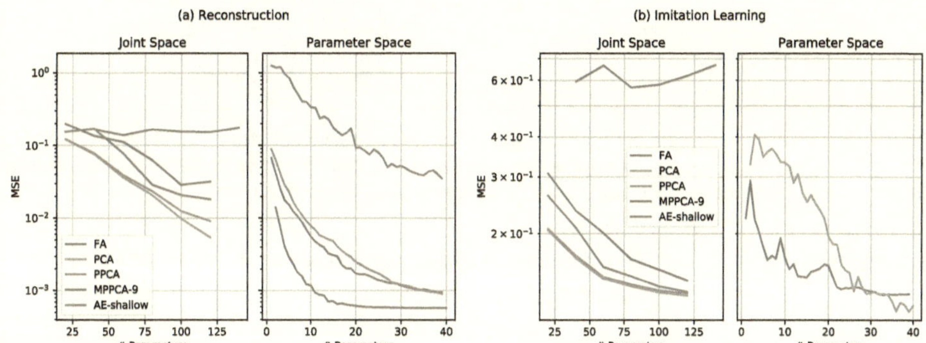

Figure 3.8.: (a) Reconstruction error on the RLBench tasks: dimensionality reduction in parameter space achieve better accuracy with less parameters. (b) Generation of most-likely movement conditioned on a given context: the dimensionality reduction in parameter space achieve similar accuracy w.r.t. in joint space, but with lower dimensionality.

terms of absolute accuracy between the joint and the parameter space (which can be imputed to an irreducible noise in the demonstrations), however, the same accuracy is reached with considerably less parameter when the dimensionality reduction is applied in the parameter space (Figure 3.8b). A more detailed analysis is presented in Appendix.

3.3.3. A Qualitative Evaluation

We want to understand the applicability of PriMos to real robotics. The study of the accuracy conducted is important, but on its own, it does not give us a feeling about how the method works in practice. For this reason, we use the 15 trajectories contained in the pouring dataset, and we measured the quantity of sugar poured in the bowl. We leave the analysis of PMMP both on real robotics to the next Chapter.

Single Movement Reconstruction. We run all the 15 trajectories reconstructed with the classic MPs and with PriMos for three times in order to average the stochasticity inherent to the experiment (a slight perturbation of the glass position or of the sugar contained in the glass might perturb the resulting quantity of sugar poured).

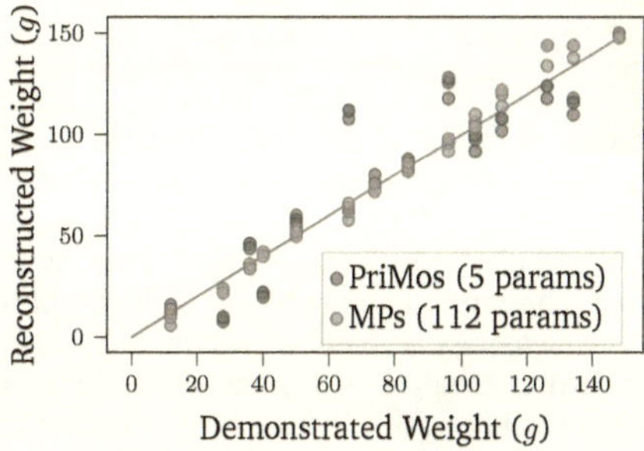

Figure 3.9.: Precision of the reconstructed weight. The diagonal line represents the best possible reconstruction. For each demonstration, we repeated three times the reconstructed movement in order.

A video of the demonstration as well of the reconstruction is available as supplementary material. Figure 3.9 represents a confusion scatter plot: on the x-axis we have the weight of the sugar observed during the demonstration, while on the y-axis we observe the reconstructed movement both with MPs (112 parameters) and with PriMos (5 parameters). The ideal situation is when the points lye down on the identity line shown

Method	0-40g	40-80g	80-120g	120-150g	Failure
Demonstrations	20%	33%	27%	20%	0%
ProMPs	43%	7%	20%	17%	13%
Pro-PriMos	53%	20%	3%	24%	0

Table 3.1.: Measurement of sugar poured from 40 movements sampled both from ProMPs and Pro-PriMos. The methods manage to roughly resemble the original distribution, however ProMPs exhibit unwanted behaviors, like pouring most of the sugar outside the bowl (classified as "failures").

in green (perfect reconstruction). We observe that PriMos, except few outliers, reaches a similar accuracy to the MPs, but with 4.5% of the parameters.

Probabilistic Representation. An important aspect of our framework is the possibility to represent the *distribution* of movements. For this reason, we show novel movements generated both with ProMPs and Pro-PriMos given 25 demonstrations. The approximation of the full covariance-matrix is usually demanding, as it scales quadratically w.r.t. the number of parameters encoding the mean movement. Figure 3.5 depicts the standard deviation of the shoulder of the robotic arm, within the demonstrated movement; Pro-PriMos seems to represent the stochasticity very similarly to the ProMPs, although the covariance matrix in the first case requires 5×5 values, in contrast to 112×112 required by the ProMPs. We also note that the variability of the Pro-PriMos seems to be lower than the ProMPs' one: this behavior is explained by the fact that for a fixed set of features, PriMos represents a subset of movements of ProMPs. We furthermore sample 40 movements both from ProMPs and Pro-PriMos with $n_c = 5$, and we measure the quantity of sugar poured. Table 3.1 shows that both the methods are able to pour all ranges of sugar (even if with different proportion from the demonstrated data). However, ProMPs occasionally fails to generate a good movement, and it pours the sugar outside the bowl. The video in the Supplementary Materials shows this situation.

3.4. Epilogue

The main contribution of this chapter is the analysis of the dimensionality reduction in parameter space in the context of MPs. Our findings suggest that this reduction is more efficient than the one in the joint space. The novel approaches (PriMos and MPPM), which operates dimensionality reduction in the parameter space using the principal component analysis, is enriched with a probabilistic treatment of the parameters in order to inherit all the convenient properties of the ProMPs. We tested our approach both on a robotic task as well as in a challenging dataset of human movements. With a mixture of principal component analyzers, we are able to further enhance the expressivity of our model and achieve higher accuracy. Our proposed methods efficiently compress the the movement primitives and allow for contextual imitation learning.

In the next chapter, we utilize the MPPMs to build an efficient reinforcement learning algorithm applicable on real robotics.

4. Off-Policy Optimization for Robotic Manipulation Skills

The research conducted in Chapter 3 suggests that the dimensionality reduction in parameter space for robotic movement is particularly efficient. We build therefore on the mixture of probabilistic principal component analyzers in the parameter space. The full probabilistic treatment allows us to condition the movement based on a context. We improve the resulting policy with a novel off-policy reinforcement gradient algorithm built on self-normalized importance sampling. Similarly to Peters et al. (2010), we introduce a trust-region based on Kullback-Leibler divergence to obtain a robust optimization.

4.1. Prologue

Learning manipulation skills is essential for enabling robots to execute a multitude of tasks, both in home and industrial environments. An important aspect of future robots is their ability to acquire, adapt or improve their skills, using demonstrations from non-expert users (Wilde et al., 2020). Arguably, despite great progress in learning to manipulate (Lillicrap et al., 2016), the acquired skills do not generalize well across different tasks and domains (Kroemer et al., 2019). Manipulation tasks are characterized by a wide variety of motions and can be related to multiple different objects (i.e., contexts). Current approaches provide manipulation policies that are restricted either to a given task or a given object and, subsequently, they do not generalize well. Indeed, even when having to execute the same task, e.g. object picking-and-placing, the performance is subject to different contextual parameters, e.g. the object's position and orientation, the sensorial feedback, and the goal position (Kupcsik et al., 2013).

Prior work on the topic has considered imitation learning (IM) of such tasks, in which a set of expert demonstrations are collected (Rueckert et al., 2015; Colomé & Torras, 2018a). Each demonstration is a set of robotic joint configurations at specific time intervals. Contexts are typically considered as vectors, describing, for example, a goal position, the position of the object to be picked, etc. However, the acquired policies are restricted by the provided data, usually overfitting expert motions, and not fulfilling the task's objective. Reinforcement learning (RL), on the other hand, holds the promise of providing control policies that will optimize over the task's goal.

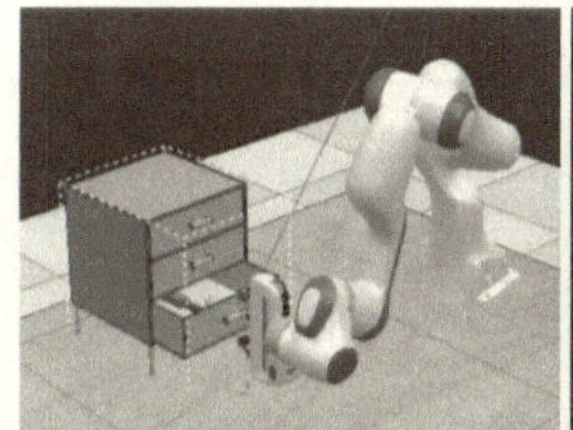

Figure 4.1.: **Left:** Simulated drawer-closing task, where the robot should close an open drawer. **Right:** Physical robot platform performing a pouring task. In this task the robot has to learn to pour the designated amount of liquid into the bowl.

Hence, researchers have opted for combining policy optimization together with the imitation-learned skills, in order to generalize over different tasks and related contexts (Cheng et al., 2018).

Among the many tools existing in literature, the framework of Probabilistic Movement Primitives (ProMPs) (Paraschos et al., 2013, 2018) enjoys favorable properties, like modulation and conditioning, which makes it a well-suited tool for imitation and improvement of robotic movements. However, the usual high-dimensional representation of robotic movements complicates the application of RL techniques, since learning on the real robot requires much more sample-efficiency than learning in simulation (Lillicrap et al., 2016). To overcome this difficulty, researchers have proposed the application of dimensionality reduction (DR) techniques in movement primitives (MPs) (Chen et al., 2015, 2016; Colomé et al., 2014; Colomé & Torras, 2014, 2018a). The resulting latent representation can help us decode the inter-dependencies between movements and contexts, allowing us to use contextual RL (Kupcsik et al., 2013) to optimize over all task-related parameters towards the achievement of the final goal.

In this chapter, we introduce a novel RL algorithm for optimizing over MPs, named LAtent-Movements Policy Optimization (LAMPO). The main focus of this algorithm is to gain sample efficiency by: 1. performing the policy updates in a reduced latent space, and 2. using off-policy gradient estimations, which re-use samples collected from previous learning episodes. In robot learning, the achieved reduced number of samples and the use of contextual policies are both important for acquiring manipulation skills through demonstrations, and, subsequently, improving them, so that those generalize across various contexts. For learning the latent representation we use Mixture of Probabilistic Principal Component Analysis (MPPCA) (Tipping & Bishop, 1999a). This method, differently from other DR techniques, is fully probabilistic, allowing us to perform conditioning. MPPCA can be seen as a Gaussian mixture model (GMM), and is, therefore, a universal density approximator, enabling the possibility of representing multimodalities and non-linear dependencies in data. Finally, MPPCA can perform data de-noising, as it is able to extract isotropic noise contained in the data.

For the off-policy estimation we use self-normalized importance sampling (Owen, 2013). Differently from other approaches (Shelton, 2001; Peshkin & Shelton, 2002), we perform a full-gradient estimation, also considering the normalization factor, which further lowers the variance, similarly to the baseline subtraction method (Deisenroth et al., 2013; Jie & Abbeel, 2010). Furthermore, we apply a forward Kullback-Leibler (KL) divergence bound between the subsequent policies, which, jointly to a KL-regularization of the context distribution, provides a robust policy optimization. We evaluated our algorithm both in simulated environments provided by RLBench (James et al., 2020) and on a real manipulator robot. The experimental analyses show that LAMPO outperforms the state of the art techniques in terms of sample efficiency on high-dimensional robotic manipulation tasks.

4.1.1. Problem Statement

We consider a context distribution $q(\mathbf{c})$ with $\mathbf{c} \in \mathbb{R}^{d_c}$, a known mapping $\mathcal{MP}(\boldsymbol{\omega})$ from a parametric space $\mathbb{R}^m$ to a robotic trajectory, an unknown probabilistic mapping $q(\boldsymbol{\omega}|\mathbf{c})$ from the context $\mathbf{c}$ to a parametric movement $\boldsymbol{\omega}$, which represents the behavior of the human demonstrator; and a user-defined reward function $R(\boldsymbol{\omega}, \mathbf{c})$. The goal is to find the policy $p_\theta(\boldsymbol{\omega}|\mathbf{c})$ that maximizes the average reward

$$\theta^* = \arg\max_\theta \int R(\boldsymbol{\omega}, \mathbf{c}) p_\theta(\boldsymbol{\omega}|\mathbf{c}) q(\mathbf{c}) \, d\mathbf{c} \, d\boldsymbol{\omega}. \tag{4.1}$$

4.1.2. Related Work

Learning from Demonstration is an emblematic paradigm for enabling robots to autonomously acquire new skills (Atkeson & Schaal, 1997).The most prominent methods in robot learning require the mapping of a set

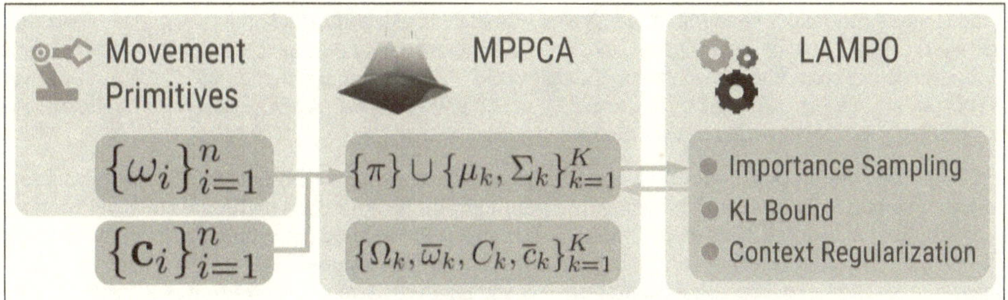

Figure 4.2.: A graphical description of our algorithm. The movement's parameters and the context are projected to a latent space using MPPCA. Our proposed off-policy reinforcement learning technique improves the latent distribution.

of demonstrated motions to a parametric model that represents dynamical systems with complex behaviors (Calinon & Billard, 2007; Ijspeert et al., 2013). Such models are generally called MPs. Dynamic MPs (DMPs) have been thoroughly explored as a robot control policy, as they are stable and robust, however, they lack generalization properties (Ude et al., 2010). ProMPs (Paraschos et al., 2013), on the other hand, is a flexible probabilistic parametric model, which represents movements as distributions over trajectories, and allows the use of probabilistic operations, like conditioning. ProMPs are able to represent complex non-linear dynamical systems, delivering impressive results in acquiring robot skills, like robot hockey (Paraschos et al., 2018).

Dimensionality reduction. Applying RL to robots is challenging due to the high-dimensional continuous spaces. A common practice is to use demonstrations to properly initialize the control policy. Even with good initialization, the performance is not optimal. Hence, researchers have opted for finding sufficient state representations, introducing DR techniques, to encode the latent features present in data (Bitzer et al., 2010). In this line of research, Colomé et al. (Colomé et al., 2014) have introduced a probabilistic DR technique of ProMPs in the joint space, using expectation-maximization (EM) over the context variables, for representing the movements in a linear low-dimensional subspace of the full configuration space. Then, the policy is further optimized via relative entropy policy search (REPS) (Peters et al., 2010). In (Colomé & Torras, 2014), the authors introduced a DR technique of the exploration parameters of DMPs when using a path integral policy improvement algorithm (Theodorou et al., 2010). In (Rueckert et al., 2015) a low dimensional latent variable model for ProMPs is extracted using fully Bayesian hierarchical models, which is used only for imitation learning. Autoencoded DMPs are proposed in (Chen et al., 2015), which uses deep autoencoders to find a latent representation of the movement from the robot's task space, towards generalizing the performance of the DMPs. In (Chen et al., 2016) the authors proposed the use of a time-dependent variational autoencoder to address the generalization challenges. Dermy et al. (Dermy et al., 2018) used auto-encoders with ProMPs for efficient human motion prediction. Colomé et al. (Colomé & Torras, 2018a) have proposed a linear DR in the joint space for learning DMPs, however, in (Colomé & Torras, 2018b), the authors proposed a reduction of the parametric space, as more appropriate for learning MPs. In (Delgado-Guerrero et al., 2020) the authors use parametric DR to learn a mapping from the MPs latent space to a reward for policy improvement.

Contextual Policy Optimization. Usually, the policies that are acquired through learning from demonstrations, are optimized in a RL setting, where the objective lies in acquiring the policy parameters for which we acquire the maximum return (Kang et al., 2018). Often the policy is warm-started by the demonstrated samples, which is further improved through policy iteration (Cheng et al., 2018), or the demonstrated and the on-policy samples are mixed (Balakrishna et al., 2020). However, the importance of generalizing skills led to the introduction of contextual RL methods (Klink et al., 2020). Indeed, contextual policy search has proven to be a data-efficient method for generalizing robot skills over contexts that represent either objectives of the robot or physical properties of the environment (Kupcsik et al., 2013; Abdolmaleki et al., 2015). The most relevant line of research to ours is the one from Colomé et al. (Colomé & Torras, 2018b), which uses contextual REPS for optimizing the policy acquired over demonstrations.

In this chapter, we propose a novel view on handling the demonstrated trajectories for acquiring lower-dimensional, non-linear latent dynamics, which can later on be optimized by LAMPO, a novel contextual policy gradient algorithm. Contrary to works in literature, our method contributes two important aspects in combining imitation learning with policy optimization. (i) Our proposed MPPCA DR technique allows us to learn a reduced latent representation of the movements over parameters and contexts, preserving their non-linear dependencies. (ii) The new contextual off-policy RL algorithm can provide gradient estimates using self-normalized importance sampling from previous experience, hence, making full use of the samples collected through different learning iterations. These advantages combined provide a framework for sample-efficient off-policy optimization of movement primitives for robot learning of high-dimensional manipulation skills.

4.2. Latent Movements Policy Optimization

Finding the optimal policy of the objective (4.1), is non-trivial. In our research problem we acquire an initial set of demonstrations, from which we learn an initialization of the parameters of the ProMPs. We start with the imitation learning part, where we aim to find an initial policy parameter set θ_0 (where the pedix 0 indicates that this set of parameters is the first of a series of improvement carried out with RL) for which $p_{\theta_0}(\omega|\mathbf{c}) \approx q(\omega|\mathbf{c})$. By maximizing the log-likelihood

$$\theta_0 = \arg\max_{\theta} \sum_{i=1}^{n} \log p_{\theta_0}(\omega_i, \mathbf{c}_i) \tag{4.2}$$

where $\omega_i, \mathbf{c}_i \sim q(\omega_i|\mathbf{c}_i)q(\mathbf{c}_i)$, we can then find the policy $p_{\theta_0}(\omega|\mathbf{c})$ by conditioning. Subsequently, we aim to use RL for policy improvement. This requires the use of a generative model of $p(\omega|\mathbf{c})$ for generating trajectories over context. At every iteration T we collect a set of rewards $\{R_i, \omega_i, \mathbf{c}_i\}_{i=Tn}^{(T+1)n}$ for n number of episodes using the Tth set of parameters θ_T and optimize objective (4.1) to acquire a better set of parameters θ_{T+1}. In our framework, as we will describe in the following, we will replace the high-dimensional ω_i with a better suited low-dimensional latent representation. Figure 4.2 presents a graphical description of the proposed algorithm, whose components and interacting part we are going to analyze in the following.

4.2.1. Imitation Learning with Dimensionality Reduction

First, let us introduce the notation and formalization of the classic ProMP framework, although any trajectory parametrization can be used. For simplicity, we will first consider only a one-dimensional trajectory $\tau^\mathsf{T} = [\tau_1, \ldots, \tau_T]^1$, where τ_i represents the joint's position at the normalized time $t_i \in [0, 1]$. Moreover, we consider n_f normalized radial-basis functions $\varphi_i : \mathbb{R} \to \mathbb{R}$ with $i \in \{1, \ldots, n_f\}$, usually, ordered to evenly cover the phase-space, centered between $[-2h, 1 + 2h]$, where h is their bandwidth. At time t the kernel features describe a n_f-dimensional column vector $\mathbf{\Phi}_t = [\varphi_1(t), \varphi_2(t), \ldots, \varphi_{n_f}(t)]^\mathsf{T}$. Assuming that the observation at

---

[1]Contrarily to (Paraschos et al., 2018), we do not consider the velocities, however, they can be incorporated without loss of generality.

time t is perturbed by a zero-mean Gaussian noise ϵ_t with variance σ_τ, we want to represent the movement $\tau_t = \Phi_t^\mathsf{T}\omega + \epsilon_t$ as a linear combination of parameters $\omega \in \mathbb{R}^{n_f}$, thus we obtain the following *maximum-likelihood estimation* (MLE) problem $\max_\omega \mathcal{L}(\omega) = \max_\omega \prod_t \mathcal{N}(\tau_t \mid \Phi_t^\mathsf{T}\omega, \sigma_t)$. The definition of MPs can be easily extended to multi-dimensional trajectories, for which at each time-step we need to represent the configuration of n_r robotic joints. In that case, the parameter vector ω lies in a m-dimensional space ($\omega \in \mathbb{R}^m$), where $m = n_f n_r$.

In our setting, each demonstration is composed of a vector of movement parameters ω and a vector representing the context $\mathbf{c}$. We want to use a mixture of probabilistic movements to find a latent space to represent jointly the movement parameters and the context,

$$
\begin{aligned}
k &\sim Cat(k|\boldsymbol{\pi}), & (k \in \{1, \ldots K\}) \\
\mathbf{z} &\sim \mathcal{N}(\mathbf{z}|\boldsymbol{\mu}_k, \boldsymbol{\Sigma}_k) & (\mathbf{z} \in \mathbb{R}^{d_z}) \\
\epsilon_\omega, \epsilon_c &\sim \mathcal{N}(\epsilon_\omega, \epsilon_c|\mathbf{0}, \mathbf{I}), & \\
\omega &= \boldsymbol{\Omega}_k \mathbf{z} + \overline{\omega}_k + \sigma_k^2 \epsilon_\omega, & (\omega \in \mathbb{R}^m) \\
\mathbf{c} &= \mathbf{C}_k \mathbf{z} + \overline{\mathbf{c}}_k + \sigma_k^2 \epsilon_c, & (\mathbf{c} \in \mathbb{R}^{d_c})
\end{aligned}
$$

where k is a categorical latent variable representing the selection of a particular Gaussian, its parameter $\boldsymbol{\pi}$ denotes the probabilities describing this selection; $\mathbf{z}$ is a Gaussian latent variable representing a specific movement-context pair in each cluster. The definition of the isotropic noises $\epsilon_\omega, \epsilon_c$ helps to compress only the useful information, and act as a denoiser in the generative model. We can group the two last equations to highlight that this setting is the same as the one described in the mixture of principal component analyzers

$$
\begin{bmatrix} \omega \\ \mathbf{c} \end{bmatrix} = \begin{bmatrix} \boldsymbol{\Omega}_k \\ \mathbf{C}_k \end{bmatrix} \mathbf{z} + \begin{bmatrix} \overline{\omega}_k \\ \overline{\mathbf{c}}_k \end{bmatrix} + \sigma_k^2 \begin{bmatrix} \epsilon_\omega \\ \epsilon_c \end{bmatrix}. \tag{4.3}
$$

Therefore, we can use the EM procedure (Tipping & Bishop, 1999a) to optimize the parameters $\{\boldsymbol{\pi}, \boldsymbol{\Omega}_k, \mathbf{C}_k, \overline{\omega}_k, \overline{\mathbf{c}}_k, \sigma_k^2\}$, assuming without loss of generality, that $\boldsymbol{\mu}_k = 0, \boldsymbol{\Sigma}_k = \mathbf{I}$. Once the maximum-likelihood parameters are obtained via EM, we are able to condition our distribution in order to generate the movement parameters ω given a context $\mathbf{c}$. To generate new movements, we first sample the latent variables $\mathbf{z}, k$ conditioned on the context, and then we establish the relation $\omega = \boldsymbol{\Omega}_k \mathbf{z} + \overline{\omega}_k$ removing the isotropic noise on the movement (as it is an unnecessary perturbation).

The generative model of $p(\mathbf{z}, k|\mathbf{c})$ results to be a GMM, $p(\mathbf{z}, k|\mathbf{c}) = p(\mathbf{z}|k, \mathbf{c})p(k|\mathbf{c})$, with the conditional responsibilities

$$
p(k|\mathbf{c}) = \frac{p(\mathbf{c}|k)\pi_k}{\sum_k p(\mathbf{c}|k)\pi_k} \tag{4.4}
$$

and $p(\mathbf{c}|k)$ is obtained marginalizing $\mathbf{z}$ from $p(\mathbf{c}|k, \mathbf{z})$ in (4.3) and

$$
p(\mathbf{z}|k, \mathbf{c}) = \mathcal{N}\left(\mathbf{z}|\mathbf{B}_k\left(\sigma_k^2 \mathbf{C}_k^\mathsf{T}(\mathbf{c} - \overline{\mathbf{c}}_k) + \boldsymbol{\Sigma}_k \boldsymbol{\mu}_k\right), \mathbf{B}_k\right) \tag{4.5}
$$

with $\mathbf{B}_k = \left(\boldsymbol{\Sigma}_k + \sigma_k^{-2} \mathbf{C}_k^\mathsf{T} \mathbf{C}_k\right)^{-1}$. In the generative model, we assume that there is no isotropic noise on the movement's parameter ω (which can be seen as noise filtering in the data),

$$
\omega = \boldsymbol{\Omega}_k \mathbf{z} + \overline{\omega}_k \quad \text{with} \quad \mathbf{z} \sim p(\mathbf{z}|k, \mathbf{c}), \quad k \sim p(k|\mathbf{c}). \tag{4.6}
$$

As already mentioned, this generative model is able to capture non-linear dependencies between the context and the robotic movement, both defining a convenient latent representation and maintaining mathematical tractability.

4.2.2. Off-Policy Reinforcement Learning

In the previous section we learned a generative model which is a mixture of probabilistic principal component analyzers. The variables $\{\mathbf{C}_k, \overline{\mathbf{c}}_k, \mathbf{\Omega}_k, \overline{\boldsymbol{\omega}}_k, \sigma_k\}_{k=1}^{K}$ describe how to project, for each probabilistic principal component analyzer k, a variable $\mathbf{z}$ to a movement-context pair. We consider that this projection should remain fixed in the RL part, while we optimize the probabilistic model of k and $\mathbf{z}$, which represents how the movements and the context are distributed in the latent space. The probabilistic model of k and $\mathbf{z}$ is described by $\{\pi\} \cup \{\boldsymbol{\mu}_k, \mathbf{\Sigma}_k\}_{k=1}^{K}$. Since $\{\mathbf{\Omega}_k, \overline{\boldsymbol{\omega}}_k, \mathbf{C}_k, \overline{\mathbf{c}}_k\}_{k=1}^{K}$ are now considered fixed, they can be considered a part of the reward signal, rather than part of the policy, changing the initial problem (4.1) into the following optimization problem

$$J(\theta) = \int R(\boldsymbol{\omega}, \mathbf{c}) p_\theta(\boldsymbol{\omega}|\mathbf{c}) q(\mathbf{c}) \, d\boldsymbol{\omega} \, d\mathbf{c} = \int \sum_{k=1}^{m} R(\mathbf{\Omega}_k \mathbf{z} + \overline{\boldsymbol{\omega}}_k, \mathbf{c}) p_\theta(\mathbf{z}, k|\mathbf{c}) q(\mathbf{c}) \, d\mathbf{c}. \tag{4.7}$$

The above formulation provides a desirable latent representation of the policy, as the latter is depended on the latent parameters $\mathbf{z}, k$ conditioned on the underlying context $\mathbf{c}$. To obtain an efficient and reliable policy optimization process, we consider the following steps. (i) At each iteration T we consider all samples collected from all past policies $\theta_1, \dots, \theta_{T-1}$. (ii) For each new set of parameters θ_T, we obtain a new conditional model $p_{\theta_T}(\boldsymbol{\omega}|\mathbf{c})$, and a new context distribution $p_{\theta_T}(\mathbf{c})$. When $p_{\theta_T}(\mathbf{c})$ diverges from $q(\mathbf{c})$, the computation of $p_{\theta_T}(\boldsymbol{\omega}|\mathbf{c})$ can suffer from numerical instabilities. To prevent such an issue, we use a KL-regularization to keep the two distributions close enough. (iii) To prevent premature convergence of the policy to a local optimum, we use a KL-constraint between the previous and the current policy distribution i.e., $\mathcal{KL}(p_\theta(\cdot|\mathbf{c})\|p_{\theta_T}(\cdot\|\mathbf{c}))$. In the following we analyze the proposed approaches for each of the above steps.

Off-Policy Estimate

Regarding the first step, we propose the use of Self-Normalized Importance Sampling (SNIS) (Owen, 2013). SNIS usually provides an estimation affected by lower variance than importance sampling, but at the price of a small bias. Its usage in the context of RL is well established as in (Shelton, 2001; Peshkin & Shelton, 2002).

Let us assume that the dataset is now composed of samples generated by T different policies parametrized with $\theta_1, \dots, \theta_T$. During the generation of the dataset, we keep information of which policy has generated which sample. Now, suppose you want to correct the distribution of a sample $\{\mathbf{z}_i, k_i, \mathbf{c}_i\}$ generated by the policy θ_ζ. Formally, θ_ζ has been chosen uniformly from $\{\theta_1, \dots, \theta_T\}$, with ζ being just a selction index, then $\mathbf{c}_i \sim p(\mathbf{z}_i, k_i|\mathbf{c}_i, \zeta) = p_{\theta_\zeta}(\mathbf{z}_i, k_i|\mathbf{c}_i)$. With the above assumptions, we can compute the importance sampling ratio

$$\rho_i = \frac{p_\theta(\mathbf{z}_i, k_i, \mathbf{c}_i, \zeta)}{p(\mathbf{z}_i, k_i, \mathbf{c}, \zeta)} = \frac{p_\theta(\mathbf{z}_i, k_i|\mathbf{c}_i) p(\zeta) q(\mathbf{c}_i)}{p(\mathbf{z}_i, k_i|\mathbf{c}_i, \zeta) p(\zeta) q(\mathbf{c}_i)} = \frac{p_\theta(\mathbf{z}_i, k_i|\mathbf{c}_i) p(\zeta) q(\mathbf{c}_i)}{p_{\theta_\zeta}(\mathbf{z}_i, k_i|\mathbf{c}_i) p(\zeta) q(\mathbf{c}_i)} = \frac{p_\theta(\mathbf{z}_i, k_i|\mathbf{c}_i)}{p_{\theta_\zeta}(\mathbf{z}_i, k_i|\mathbf{c}_i)}$$

where $p(\zeta) = 1/T$, then (4.7) yields to[2]

$$J(\theta) \approx \hat{J}(\theta) = \sum_{i=1}^{n} \frac{\rho_i}{\nu} R_i \tag{4.8}$$

with $\nu = \sum_i \rho_i$.

---

[2]The dependencies of ν and ρ_i from θ are omitted for simplicity, however this dependency should be considered when computing the gradient.

Context Regularization

For the second step, we introduce a regularization over the approximation of the contextual distribution. When we condition our mixture model to compute $p(\mathbf{z}, k|\mathbf{c})$, we need the ratio $p(\mathbf{z}, k, \mathbf{c})/p(\mathbf{c})$. For simplicity of notation we return back to the objective (4.7), which for this step can be rewritten as

$$\int \sum_{k=1}^{K} R(\boldsymbol{\Omega}_k \mathbf{z} + \overline{\boldsymbol{\omega}}_k, \mathbf{c}) \frac{p_\theta(\mathbf{z}, k, \mathbf{c})}{p_\theta(\mathbf{c})} q(\mathbf{c}) \, d\mathbf{c}. \tag{4.9}$$

After the imitation learning phase we acquire an approximation of the context distribution $p_\theta(\mathbf{c}) \approx q(\mathbf{c})$. However, during the optimization of θ, the distribution p_θ eventually changes, thus, affecting the contextual distribution approximation. To avoid the high variance of this process, we encourage $p_\theta(\mathbf{c})$ to stay close to the initial approximation of $q(\mathbf{c})$ using the KL-divergence, i.e. close to $p_{\theta_0}(\mathbf{c})$, where θ_0 are the parameters of the first policy, acquired immediately after the imitation learning. Since $p_\theta(\mathbf{c})$ is a mixture of Gaussians, we cannot compute in closed form the KL between $p_\theta(\mathbf{c})$ and $p_{\theta_0}(\mathbf{c})$. To overcome this issue, we compute the KL between $p_\theta(\mathbf{c}, k)$ and $p_{\theta_0}(\mathbf{c}, k)$, which is an upper-bound of the previous term

$$\int \sum_k p_\theta(\mathbf{c}, k) \log \frac{p_\theta(\mathbf{c}, k)}{p_{\theta_0}(\mathbf{c}, k)} \, d\mathbf{c} = \int p(\mathbf{c}) \underbrace{\sum_k p_\theta(k|\mathbf{c}) \log \frac{p_\theta(k|\mathbf{c})}{p_{\theta_0}(k|\mathbf{c})} \, d\mathbf{c}}_{\text{Always non-negative}} + \int p_\theta(\mathbf{c}) \log \frac{p_\theta(\mathbf{c})}{p_{\theta_0}(\mathbf{c})} \, d\mathbf{c}.$$

The analytical expression of the KL-divergence between $p_\theta(\mathbf{c}, k)$ and $p_{\theta_0}(\mathbf{c}, k)$ takes the form of

$$
\begin{aligned}
\eta_\theta &= \mathcal{KL}\left(p_\theta(\mathbf{c}, k) \| p_{\theta_0}(\mathbf{c}, k)\right) \\
&= \sum_k p_\theta(k) \Big(\mathcal{H}\left(p_\theta(\mathbf{c}|k), p_{\theta_0}(\mathbf{c}|k)\right) - \mathcal{H}\left(p_\theta(\mathbf{c}|k), p_{\theta_0}(\mathbf{c}|k)\right) \Big) + \mathcal{H}\left(p_\theta(k), p_{\theta_0}(k)\right) - \mathcal{H}\left(p_\theta(k), p_\theta(k)\right)
\end{aligned} \tag{4.10}
$$

where $\mathcal{H}$ is the entropy, and since $p_\theta(\mathbf{c})$ and $p_{\theta_0}(\mathbf{c})$ are Gaussian distributions, and $p_\theta(k)$ and $p_{\theta_0}(k)$ are categorical distributions, (4.10) is computable in closed form. Using the divergence in (4.10) as a regularization term in our objective (4.8) stabilizes the optimization process.

Trust Region

In order to have a robust optimization process, it is a common practice to introduce a KL-constraint (or regularization) between two subsequent policies during the optimization (Peters et al., 2010; Schulman et al., 2015). These methods avoid the premature convergence of the policy to a local optimum and they guarantee a smooth optimization process. Usually, the KL-constraint is imposed by keeping the policy close to the samples generated by the previous policy (Peters & Schaal, 2008). In our work, however, the samples are originated from all previous policies, making the above approach not applicable. We, instead, want to keep our policy close only to the previous one. Furthermore, in our case it is possible to compute the KL in a closed form. Although the KL is not computable in a closed form when using GMMs, considering the latent variable k jointly with $\mathbf{z}$ makes such a computation tractable, as shown in the previous subsection. Hence, the KL divergence between the optimized policy and the policy at iteration $T - 1$, can be computed as

$$g_\theta(\mathbf{c}) = \mathcal{KL}\left(p_{\theta_{T-1}}(\cdot|\mathbf{c}) \| p_\theta(\cdot|\mathbf{c})\right). \tag{4.11}$$

4.2.3. Off-policy Improvement

Until this point, we have formalized the optimization objective for our RL algorithm, introducing the off-policy objective, the context and the the trust region regularization. Note that all these quantities are differentiable w.r.t. the policy parameters. We can, therefore, formulate the gradient estimation of the policy optimization problem. To ensure the parameters consistency, we encode the categorical distribution as $\pi = e^{\theta_i} / \sum_{i=1}^{K} e^{\theta_i}$, and the covariances Σ_k as positive-definite diagonal matrices. By combining (4.8) (4.10) and (4.11), the policy improvement can be framed as a constrained optimization problem

$$\max_{\theta} \hat{J}(\theta) - \gamma\eta_\theta$$

$$\text{s.t. } n^{-1} \sum_{i=1}^{n} g_\theta(\mathbf{c}_i) \leq \chi \tag{4.12}$$

where χ is the upper-bound of the forward KL, and γ is the regularization constant which controls how strongly the regularizator η_θ defined in Equation 4.10 impacts the objective. Since we have access to the analytical gradient of all the entities composing the optimization problem, we can use any constraint-optimization solver; we have chosen the Sequential Least SQuare Program (SLSQP). The gradient w.r.t. the approximated objective (4.8) is

$$\nabla_\theta \hat{J}(\theta) = \nu^{-1} \sum_{i=1}^{n} \left(\nabla_\theta \rho_i - \nu^{-1} \rho_i \nabla_\theta \nu \right) R_i$$

$$= \sum_{i=1}^{n} \frac{\rho_{i,j}}{\nu} \nabla_\theta \log p_\theta(\mathbf{z}_i, k_i | \mathbf{c}_i) \left(R_i - \hat{J}(\theta) \right), \tag{4.13}$$

which turns out to be equal to the SNIS of the classic gradient with baseline subtraction, which leads to lower variance in the gradient estimation (Williams, 1992; Sutton et al., 2000; Deisenroth et al., 2013). We could compute $\nabla_\theta \log p_\theta(\mathbf{z}_i, k_i | \mathbf{c}_i)$ by taking the gradient of (4.6), but due to the matrix inversion required for computing $\mathbf{B}_k$, the computation presents numerical instability. We propose, instead, the derivation of the gradient from

$$\log p_\theta(\mathbf{z}, k | \mathbf{c}) = \log p_\theta(\mathbf{c} | \mathbf{z}, k) + \log p_\theta(\mathbf{z} | k) + \log \pi_k - \log \underbrace{\sum_{j} \pi_j \int p_\theta(\mathbf{c} | \mathbf{z}, j) p_\theta(\mathbf{z} | j) \, d\mathbf{z}}_{p_\theta(\mathbf{c} | j)} \tag{4.14}$$

which is solvable in closed form, where $p_\theta(\mathbf{c} | j) = \mathcal{N}(\mathbf{c} | \mathbf{C}_j \boldsymbol{\mu}_j + \overline{\mathbf{c}}_j, \sigma_j^2 \mathbf{I} + \mathbf{C}_j \Sigma_j \mathbf{C}_j^{-1})$, does not require the inversion of $\mathbf{B}_k$. We do not show the complete computation of the gradient of optimization problem due to space limitation, but it can be computed either manually via simple algebraic manipulation, or via automatic differentiation tools, like `PyTorch` or `Tensorflow`[3]. An overall description of the steps required in LAMPO are listed in Algorithm 1.

4.3. Empirical Analysis

For evaluating the performance of LAMPO, we study the effect of motion multimodalities, comparing against baseline approaches. We are also testing whether LAMPO scales to different contexts and dimensionalities

---

[3]In this work we use PyTorch.

Algorithm 1 LAtent Movement Policy Optimization

1: **input:** Dataset $\{\tau_i, c_i\}_{i=1}^{n_d}$ of trajectories and contexts.
2: **for each** trajectory τ_i compute the parameter vector ω_i.
3: EM to find the MPPCA's parameters $\pi, \Omega_k, \overline{\omega_k}, C_k, \overline{c}_k$ (Tipping & Bishop, 1999a)
4: assuming $\mu_k = 0, \Sigma_k = \mathbb{I}$.
5: Set $\theta_0 \equiv \{\pi\} \cup \{\mu_k, \Sigma_k\}_{k=1}^K$.
6: **for** $T \in \{1 \cdots N_T\}$ **do**
7:     **for** $i \in \{n(T-1) \ldots nT\}$ **do**
8:         Observe context c_i
9:         Perform movement ω_i with (4.6)
10:         collect reward R_i
11:     **end for**
12:     Compute a differentiable model of $\hat{J}(\theta)$, η_θ and $g(\theta)$ using (4.8), (4.10) and (4.11).
13:     Solve (4.12) for $\{\pi\} \cup \{\mu_k, \Sigma_k\}_{k=1}^K$ using SLSQP.
14:     Update $\theta_T \equiv \{\pi\} \cup \{\mu_k, \Sigma_k\}_{k=1}^K$.
15: **end for**

---

of a given problem. Finally, we train and validate our proposed method on a real robotic platform. In the following, we provide a description and analysis of the conducted experiments, proving that our novel method efficiently learns complex skills on a real robot.

4.3.1. Experimental Setup

For the evaluation of the proposed algorithm, we have conducted a series of experiments both in virtual environments and on a real manipulator robot. In what follows, we describe in detail the experimental setup of these evaluations.

Virtual environments.

2d-reacher is a toy simulated environment where a robotic manipulator is composed of two revolute joints, and two links of length one. The task consists in reaching four different target areas, as depicted in Figure 4.3-left. The user can select the number of clusters from which one can generate the respective goals, to acquire diverse and of different complexity movements in the dataset.

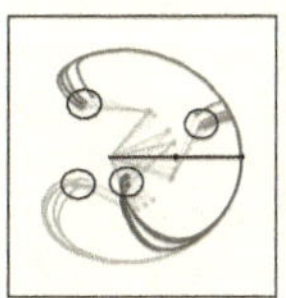

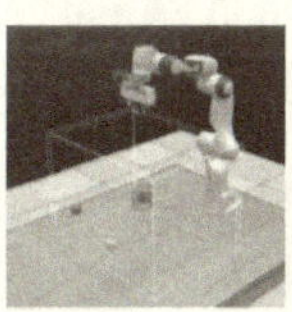

Figure 4.3.: **Left:** Our 2d-reacher.The goal is to reach some given point.Tthe number of clusters of the goal-positions can be modified to generate datasets with different degrees of non-linearities. **Center:** The reach-target from RLBench. **Right:** Our real-robot reacher.

The movements are encoded by 20 radial basis functions, for a total of 40 parameters. We add to this set an extra-parameter to encode the duration of the movement (we will keep a similar setting in all the experiments).

reach-target, defined in the RLBench suite (James et al., 2020). It comprises of a 6DoF robotic arm equipped with a simple gripper. The goal is to reach an object in the scene (Figure 4.3-center, Table 4.1). Note that, to the best of our knowledge, we are the first to conduct experiments on the RLBench suite, hence there are no comparative results in literature.

close-drawer, utilizes the same robot as above, and consists of closing a drawer that appears in the scene in different positions and with different orientations (see Figure 4.1-left, Table 4.1). The context vector has 94 dimensions. Note the high-dimensionality of this task, keeping in mind that RLBench provides us only

task	dof	context	parameters	reward
`2d-reacher`	2	goal position, 2d	40	euclidean distance
`reach-target`	6	goal position, 3d	141	1 if success, else -euclidean distance
`close-drawer`	6	goal position, 94d	141	1 if success, else -euclidean distance
`rr-reacher`	7	goal position, 3d	141	euclidean distance
`rr-pouring`	7	goal weight, 1d	141	absolute difference

Table 4.1.: Experimental details. Notice that the `close-drawer` has a high-dimensional context.

with 200 demonstrations. All reported results for the simulation environments have been averaged over 10 different experiments.

Real robot experiments.
`rr-reacher` is learned and performed on a real 7dof robot. The goal is to reach the position of a marker on a table (Figure 4.3-right, Table 4.1).

`rr-pouring` is defined in the same robot as before. We execute a task where the goal is to pour some granular material from a glass into a bowl, till the desired amount of material is poured. Here, the context is 1d, i.e., the quantity of the material, measured by a digital scale placed under the bowl (see Figure 4.1-right, Table 4.1). This task, although lower-dimensional than the previous one, is harder to learn, as there is some stochasticity perturbing the experiments, e.g., small variations of the sugar quantity from episode to episode.

4.3.2. Evaluation in Virtual Environments

For providing insights regarding the performance of LAMPO, we have compared it against the most recent state-of-the-art algorithm by Colomé & Torras (2018b), that uses GMMs in the parameter space of ProMPs, after performing DR, further optimizing the obtained policies through imitation learning using REPS. Specifically, due to the assumptions done by Colomé & Torras (2018b) in their algorithmic solution, we decided, first, to compare on the toy experiment of `2d-reacher`. First, we evaluate the selection of different number of clusters and latent space dimensions used for the MMPCA and GMMs approaches in LAMPO and CT respectively, in imitation learning. Figure 4.6 shows the heatmaps of the ablation study conducted over the two algorithms, for different numbers of clusters and latent dimensions. Note that the result depicts the performance of the obtained rewards in `2d-reacher`, i.e. Euclidean distance to the goal. It is evident that CT algorithm cannot decode different clusters of motions when compared to our approach. We have also experimented on the policy improvement setting, depicted in Figure 4.4. In this evaluation, we keep the latent space dimension fixed for CT and LAMPO, but we test the performance for different number of goal-clusters (from 1 to 4). For a fair comparison, we also have included a second baseline method, i.e. a GMM+REPS approach, in which we do not conduct any DR method on the collected data. As expected, while for one category of movements ($K = 1$) LAMPO and CT have similar performance, when we decode the movement in different clusters, CT performance is suboptimal. Moreover, the simple GMM+REPS approach performs poorly in all cases, showing the importance of acquiring a low-dimensional latent representation that preserves the motion's non-linearities.

After the above observations, we continue evaluating LAMPO on high dimensional problems. Starting with the `reach-target` task, we ablated the performance of LAMPO and CT for the RL setting (Figure 4.7).

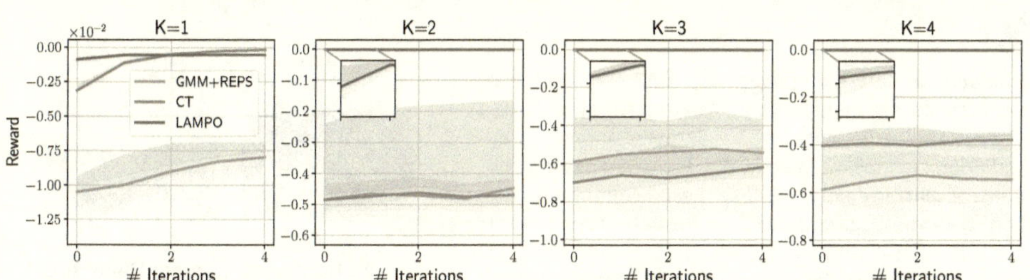

Figure 4.4.: Policy improvement performance of LAMPO against the baseline method CT and GMM+REPS in the `reacher-2d` task, using a different number of clusters K. CT struggles to learn at the presence of more than one clusters, while GMM+REPS performs poorly at all cases.

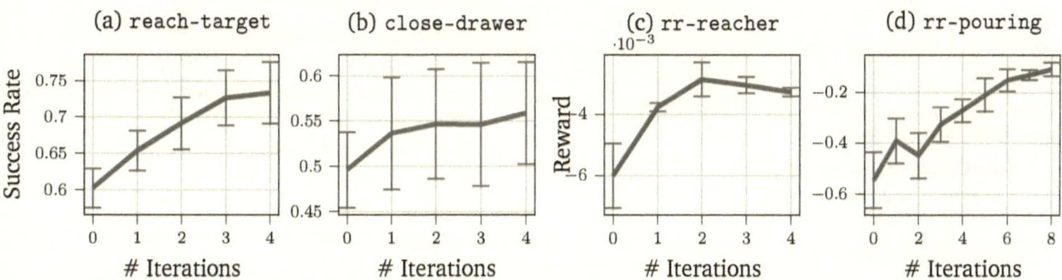

Figure 4.5.: Learning curve accompanied with 95% confidence intervals on simulated and real robotic environments. LAMPO shows often a substantial improvement over the imitation-learning phase.

In Figure 4.7a we compare both algorithms for different initial number of demonstrations in the imitation learning phase. The light-green and light–blue curves indicate performance for the IM policy, while the red and blue the curves for the RL policy for CT and LAMPO respectively. Moreover, we tested the influence of different batch sizes during the RL phase, while using the same number of samples for imitation learning (Figure 4.7b). LAMPO outperforms the baseline in both cases, when CT struggles with the multimodality of the movements in the respective task. We can see that after collecting only 200 samples during RL, the performance of LAMPO stabilizes, meaning that the collection of more samples is unnecessary. This observation serves as an empirical proof for the sample-efficiency of LAMPO, thanks to his novel off-policy sample re-use.

From the results of the previous comparative study, we have concluded on a hyper-parameter setting for LAMPO, which can provide improved policies in a data-efficient way. For further evaluation, we also conducted experiments of LAMPO in another simulated high-dimensional problem, the `close-drawer` (Figure 4.5b). While we do not achieve the same success rate as in `reach-target` in Fig. 4.5a, we can still see how LAMPO ameliorates the initial poor performance obtained by demonstrations. In the future, we want to study more expressive models for learning MPs combined with LAMPO, to cover the whole space of solutions. Note that we have not included the baseline in this experiment, since from the results obtained in the `reach-target` task, it is evident that they cannot encode the multimodality of movements in complex tasks.

4.3.3. Validation on the Real Robot

To further validate the efficieny of LAMPO, we tested it on a real robotic manipulator. As we have already described in subsection 4.3.1, we have trained and validated LAMPO on two tasks. Specifically, Fig. 4.5c depicts the results of learning with LAMPO the `rr-reacher` task, where we can observe the consistent improvement during the RL phase. Indeed, we end up with a policy that achieves significantly low errors (order of magnitude less than 10^{-2}). Note that there is also a consistent error in the measurement of the fingertip and the target position from the motion capture system, e.g. due to possible occlusions or the nominal system noise.

Fig. 4.5d shows the results obtained for the `rr-pouring` task. We can see an impressive slope in the learning curve, that starts from a low performance after imitation learning, achieving optimal results after the LAMPO policy improvement with the off-policy sample re-use. Note that, also, in this case the results are dependent on the digital scale nominal error ($\sim 2\mathrm{gr}$) and other sources of noise.

4.4. Epilogue

This chapter introduces LAMPO, a novel contextual RL algorithm for optimizing over MPs while learning high-dimensional robotic manipulation tasks with different contexts. The main advantage of LAMPO is twofold; first, it offers a reduced latent representation of non-linear dynamics, while encoding the dependencies between movements and task-related contexts. Second, it gains in sample-efficiency with off-policy gradient estimations, re-using samples collected from previous learning episodes. For learning the latent representation we use MPPCA, which is fully probabilistic, and allows us to perform conditioning on the task contexts. For the off-policy estimation, we use SNIS combined with a full-gradient estimate, also considering the normalization factor, which further lowers the estimation variance. Our contextual RL method uses a forward KL-divergence on the policy improvement, which, jointly with a KL-regularization of the context distribution, provides a robust policy optimization. We evaluated our algorithm both on simulated environments and on a real manipulator robot. The experimental analyses show that LAMPO outperforms the state of the art techniques in terms of sample efficiency on high-dimensional robotic manipulation tasks. In the future, we will explore the possibility of introducing a Dirichlet prior to the mixture's weight in order to stabilize the learning, as well as a modification to the EM algorithm to incorporate the effect on the configuration space.

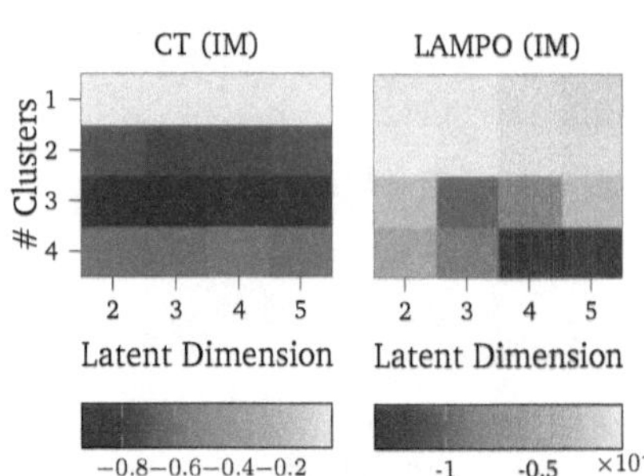

Figure 4.6.: Heatmaps of the rewards obtained in the environment `2d-reacher` for the imitation learning policies (IM) for LAMPO and CT (Colomé & Torras, 2018b), across different numbers of clusters and latent space dimensions. Our algorithm outperformns CT especially at the presence of different clusters of movements. Note the different color scale.

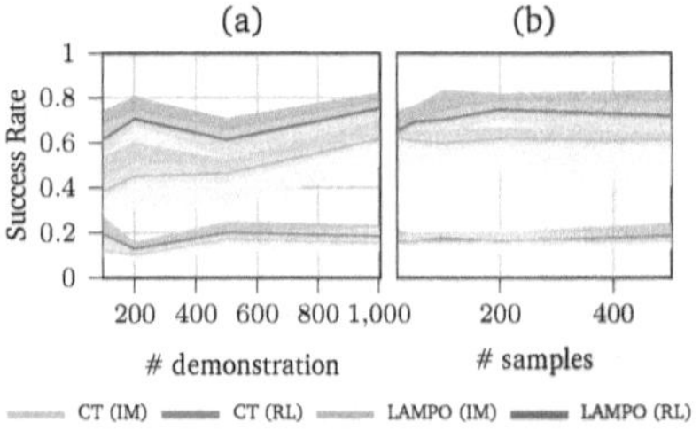

Figure 4.7.: `reach-target`: (a) While keeping the number of samples per iteration fixed to 500, we use different number of demonstrations. (b) While keeping the number of demonstrations fixed to 1000, we use different amount of samples per policy improvement.

5. Offline Reinforcement Learning with a Nonparametric Off-Policy Policy Gradient

In the previous chapter, we analyzed a novel episodic off-policy algorithm. Episodic reinforcement learning does not scale with the number of parameters and does not exploit the state-action formulation. In this chapter, we propose a novel step-based off-policy algorithm capable of working in an offline scenario. State-of-the-art approaches either omit a term in the gradient expansion introducing a highly biased estimate or rely on a high variance importance sampling correction. In contrast, our method builds a closed-form solution of a nonparametric Bellman equation that allows a full gradient expansion, delivering better estimates in terms of bias/variance trade-off.

5.1. Prologue

Reinforcement learning has made overwhelming progress in recent years, especially when applied to board and computer games, or simulated tasks (Mnih et al., 2015; Haarnoja et al., 2018; Schulman et al., 2015). However, in comparison, only a little improvement has been achieved on real-world tasks. One of the reasons of this gap is that the vast majority of reinforcement learning approaches are on-policy. On-policy algorithms require that the samples are collected using the optimization policy; and therefore this implies that a) there is little control on the environment and b) samples must be discarded after each policy improvement, causing high sample inefficiency. In contrast, off-policy techniques are theoretically more sample efficient, because they decouple the procedures of data acquisition and policy update, allowing for the possibility of sample-reuse, and enable a higher degree of control on the data-acquisition process, which allows for safe interaction. These two properties are of crucial importance for real-world scenarios. However, classical off-policy algorithms like Q-learning with function approximation and fitted Q-iteration (Ernst et al., 2005; Riedmiller, 2005) are not guaranteed to converge (Baird, 1995; Lu et al., 2018), and allow only discrete actions. More recent semi-gradient[1] off-policy techniques, like Off-PAC (Degris et al., 2012b) and DDPG (Silver et al., 2014; Lillicrap et al., 2016) often perform sub-optimally, especially when the collected data is strongly off-policy, due to the biased semi-gradient update (Fujimoto et al., 2019). Off-policy algorithms based on importance sampling (Shelton, 2001; Meuleau et al., 2001; Peshkin & Shelton, 2002) deliver an unbiased estimate of the gradient but suffer from high variance and are generally only applicable with stochastic policies. Moreover, they require the full knowledge of the behavioral policy, making them unsuitable when data stems from a human demonstrator. Additionally, model-based approaches like PILCO (Deisenroth & Rasmussen, 2011) may be considered to be off-policy. Such probabilistic nonlinear trajectory optimizers are limited to the finite-horizon domain and suffer from coarse approximations when propagating the state distribution through time. To address all previously highlighted issues in state-of-the-art off-policy approaches, we propose a new algorithm, the nonparametric off-policy policy gradient (NOPG) (Tosatto et al., 2020b), a full-gradient estimate based on

---

[1]We adopt the terminology from (Imani et al., 2018).

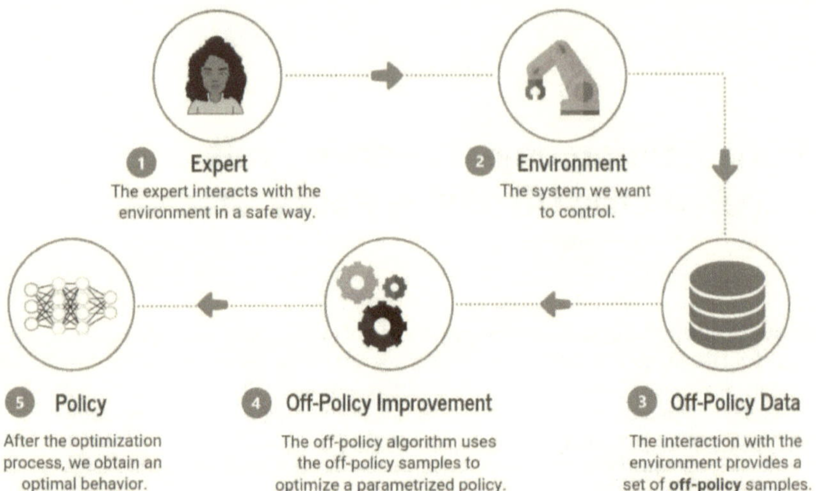

Figure 5.1.: In the off-policy reinforcement learning scheme, the policy can be optimized using an off-policy dataset. This allows for safer interaction with the system and for better sample efficiency.

the closed-form solution of a nonparametric Bellman equation. We avoid the high variance of importance sampling techniques and allow for the use of human demonstrations, and unlike other nonparametric methods like PILCO, our approach allows for multimodal state-transitions, and can handle the infinite-horizon setting. Figure 5.1 shows the optimization cycle of NOPG. A behavioral policy, represented by a human demonstrator, provides (possibly suboptimal) trajectories that solve a task. NOPG optimizes a policy from off-line and off-policy samples. The two other approaches, semi-gradient and path-wise importance sampling, do not work in this scenario.

In this chapter, we present both the theoretical foundations of our approach, and an empirical analysis to compare the quality of our gradient estimate and the sample efficiency w.r.t. state-of-the art techniques.

5.1.1. Problem Statement

Consider the reinforcement learning problem of an agent interacting with a given environment, as abstracted by a Markov decision process (MDP) and defined over the tuple $(\mathcal{S}, \mathcal{A}, \gamma, P, R, \mu_0)$ where $\mathcal{S} \equiv \mathbb{R}^{d_s}$ is the state space, $\mathcal{A} \equiv \mathbb{R}^{d_a}$ the action space, the transition-based discount factor γ is a stochastic mapping between $\mathcal{S} \times \mathcal{A} \times \mathcal{S}$ to $[0, 1)$, which allows for unification of episodic and continuing tasks (White, 2017), offering, among others, a natural representation of task termination (where $\gamma = 0$). To keep however the theory simple, we will assume that $\gamma(\mathbf{s}, \mathbf{a}, \mathbf{s}') \leq \gamma_c$ where $\gamma_c < 1$. The transition probability from a state $\mathbf{s}$ to $\mathbf{s}'$ given an action $\mathbf{a}$ is governed by the conditional density $P(\mathbf{s}'|\mathbf{s}, \mathbf{a})$. The stochastic reward signal R for a transition $(\mathbf{s}, \mathbf{a}, \mathbf{s}') \in \mathcal{S} \times \mathcal{A} \times \mathcal{S}$ is drawn from a distribution $R(\mathbf{s}, \mathbf{a}, \mathbf{s}')$ with mean value $\mathbb{E}_{\mathbf{s}'}[R(\mathbf{s}, \mathbf{a}, \mathbf{s}')] = r(\mathbf{s}, \mathbf{a})$. The initial distribution $\mu_0(\mathbf{s})$ denotes the probability of the state $\mathbf{s} \in \mathcal{S}$ to be a starting state. A policy π is a stochastic or deterministic mapping from $\mathcal{S}$ onto $\mathcal{A}$, usually parametrized by a set of parameters θ.

We define an *episode* as $\tau \equiv \{\mathbf{s}_t, \mathbf{a}_t, r_t, \gamma_t\}_{t=1}^{\infty}$ where

$$\mathbf{s}_0 \sim \mu_0(\cdot); \quad \mathbf{a}_t \sim \pi(\cdot \mid \mathbf{s}_t); \quad \mathbf{s}_{t+1} \sim p(\cdot \mid \mathbf{s}_t, \mathbf{a}_t); \quad r_t \sim R(\mathbf{s}_t, \mathbf{a}_t, \mathbf{s}_{t+1}); \quad \gamma_t \sim \gamma(\mathbf{s}_t, \mathbf{a}_t, \mathbf{s}_{t+1}).$$

In this chapter we consider the discounted infinite-horizon setting, where the objective is to maximize the expected return

$$J_\pi = \mathop{\mathbb{E}}_{\tau}\left[\sum_{t=0}^{\infty} r_t \prod_{i=0}^{t} \gamma_i\right]. \tag{5.1}$$

It is propaedeutic to introduce two important quantities: the *stationary state visitation* μ_π and the *value function* V_π. We naturally extend the stationary state visitation defined by (Sutton et al., 2000) with the transition-based discount factor

$$\mu(\mathbf{s}) = \mathop{\mathbb{E}}_{\tau}\left[\sum_{t=0}^{\infty} p(\mathbf{s} = \mathbf{s}_t | \pi, \mu_0) \prod_{i=1}^{t} \gamma_i\right],$$

or, equivalently, as the fixed point of

$$\mu(\mathbf{s}) = \mu_0(\mathbf{s}) + \int_{\mathcal{S}} \int_{\mathcal{A}} p_\gamma(\mathbf{s}|\mathbf{s}', \mathbf{a}')\pi(\mathbf{a}'|\mathbf{s}')\mu_\pi(\mathbf{s}')\, \mathrm{d}\mathbf{s}'\, \mathrm{d}\mathbf{a}'$$

where, from now on, $p_\gamma(\mathbf{s}'|\mathbf{s}, \mathbf{a}) = p(\mathbf{s}'|\mathbf{s}, \mathbf{a})\mathbb{E}_{\gamma \sim \gamma(\mathbf{s}, \mathbf{a}, \mathbf{s}')}[\gamma]$. The value function

$$V_\pi(\mathbf{s}) = \mathop{\mathbb{E}}_{\tau}\left[\sum_{t=0}^{\infty} r_t \prod_{i=0}^{t} \gamma_i \Big| \mathbf{s}_0 = \mathbf{s}\right],$$

corresponds to the fixed point of the Bellman equation,

$$V_\pi(\mathbf{s}) = \int_{\mathcal{A}} \pi(\mathbf{a}|\mathbf{s})\left(r(\mathbf{s}, \mathbf{a}) + \int_{\mathcal{S}} V_\pi(\mathbf{s}')p_\gamma(\mathbf{s}'|\mathbf{s}, \mathbf{a})\, \mathrm{d}\mathbf{s}'\right)\mathrm{d}\mathbf{a}.$$

The state-action value function is defined as

$$Q_\pi(\mathbf{s}, \mathbf{a}) = r(\mathbf{s}, \mathbf{a}) + \int_{\mathcal{S}} V_\pi(\mathbf{s}')p_\gamma(\mathbf{s}'|\mathbf{s}, \mathbf{a})\, \mathrm{d}\mathbf{s}'.$$

The expected return (5.1) can be formulated as

$$J_\pi = \int_{\mathcal{S}} \mu_0(\mathbf{s})V_\pi(\mathbf{s})\, \mathrm{d}\mathbf{s} = \int_{\mathcal{S}} \int_{\mathcal{A}} \mu_\pi(\mathbf{s})\pi(\mathbf{a}|\mathbf{s})r(\mathbf{s}, \mathbf{a})\, \mathrm{d}\mathbf{a}\, \mathrm{d}\mathbf{s}.$$

Policy Gradient Theorem. The reinforcement learning objective (5.1) can be maximized via gradient ascent. The gradient of J_π w.r.t. the policy parameters θ is

$$\nabla_\theta J_\pi = \int_{\mathcal{S}} \int_{\mathcal{A}} \mu_\pi(\mathbf{s})\pi_\theta(\mathbf{a}|\mathbf{s})Q_\pi(\mathbf{s}, \mathbf{a})\nabla_\theta \log \pi_\theta(\mathbf{a}|\mathbf{s})\, \mathrm{d}\mathbf{a}\, \mathrm{d}\mathbf{s},$$

as stated in the policy gradient theorem (Sutton et al., 2000). When it is possible to interact with the environment with the policy π_θ, one can approximate the integral by considering the state-action as a distribution (up to a normalization factor) and use the samples to perform a Monte-Carlo (MC) estimation (Williams, 1992).

The Q-function can be estimated via Monte-Carlo sampling, approximate dynamic programming or by direct Bellman minimization. In the off-policy setting, we do not have access to the state-visitation μ_π induced by the policy, but instead we observe a different state distribution. While estimating the Q-function with the new state distribution is well established in the literature (Watkins & Dayan, 1992; Ernst et al., 2005), the shift in the state visitation $\mu_\pi(\mathbf{s})$ is more difficult to obtain. State-of-the-art techniques either omit to consider this shift (we refer to these algorithms as *semi-gradient*), or they try to estimate it via importance sampling correction. These approaches will be discussed in detail in Section 5.1.2.

5.1.2. Related Work

Off-policy policy gradient estimation can be divided in three different techniques: semi-gradient approaches, importance sampling correction, and model based estimation. Semi-gradient approaches omit one term in the gradient computation, which causes an estimation bias (Imani et al., 2018). The importance sampling correction, although unbiased, suffers from high variance, which makes it often unpractical (Owen, 2013). Model based approaches rely on a model's estimation, and optimize the policy following this model. However, the model's error propagates in the number of steps, adding also a significant source of bias (Deisenroth & Rasmussen, 2011).

Semi-Gradient Approaches

The off-policy policy gradient theorem was the first proposed off-policy actor-critic algorithm (Degris et al., 2012b). Since then, it has been used by the vast majority of state-of-the-art off-policy algorithms (Silver et al., 2014; Lillicrap et al., 2016; Schulman et al., 2015, 2017; Haarnoja et al., 2018). Nonetheless, it is important to note that this theorem and its successors, introduce two approximations to the original policy gradient theorem (Sutton et al., 2000). First, semi-gradient approaches consider a modified discounted infinite-horizon return objective $\hat{J}_\pi = \int \rho_\beta(\mathbf{s})V_\pi(\mathbf{s})\,\mathrm{d}\mathbf{s}$, where $\rho_\beta(\mathbf{s})$ is state distribution under the behavioral policy π_β. Second, the gradient estimate is modified to be

$$
\begin{aligned}
\frac{\partial}{\partial \theta}\hat{J}_\pi &= \frac{\partial}{\partial \theta}\int_{\mathcal{S}} \rho_\beta(\mathbf{s})V_\pi(\mathbf{s})\,\mathrm{d}\mathbf{s} \\
&= \frac{\partial}{\partial \theta}\int_{\mathcal{S}} \rho_\beta(\mathbf{s})\int_{\mathcal{A}} \pi_\theta(\mathbf{a}|\mathbf{s})Q_\pi(\mathbf{s},\mathbf{a})\,\mathrm{d}\mathbf{a}\,\mathrm{d}\mathbf{s} \\
&= \int_{\mathcal{S}} \rho_\beta(\mathbf{s})\int_{\mathcal{A}} \underbrace{\frac{\partial}{\partial \theta}\pi_\theta(\mathbf{a}|\mathbf{s})Q_\pi(\mathbf{s},\mathbf{a})}_{\text{A}} + \underbrace{\pi_\theta(\mathbf{a}|\mathbf{s})\frac{\partial}{\partial \theta}Q_\pi(\mathbf{s},\mathbf{a})}_{\text{B}}\,\mathrm{d}\mathbf{a}\,\mathrm{d}\mathbf{s} \qquad (5.2) \\
&\approx \int_{\mathcal{S}} \rho_\beta(\mathbf{s})\int_{\mathcal{A}} \frac{\partial}{\partial \theta}\pi_\theta(\mathbf{a}|\mathbf{s})Q_\pi(\mathbf{s},\mathbf{a})\,\mathrm{d}\mathbf{a}\,\mathrm{d}\mathbf{s},
\end{aligned}
$$

where the term B related to the derivative of Q_π is ignored. The authors provide a proof that this biased gradient, or *semi-gradient*, still converges to the optimal policy in a tabular setting (Degris et al., 2012b; Imani et al., 2018). However, further approximation (e.g., given by the critic and by finite sample size), might disallow the convergence to a satisfactory solution. It might be deceiving to think that these algorithms are in fact off-policy: although they work correctly sampling from the replay memory (which discards *the oldest* samples), they have shown to fail with samples generated via a completely different process (Imani et al., 2018; Fujimoto et al., 2019). For this reason, we don't consider semi-gradient approaches to be promising for off-policy optimization.

Importance Sampling Approaches

One way to obtain an unbiased estimate of the policy gradient in an off-policy scenario is to re-weight every trajectory via importance sampling (Meuleau et al., 2001; Shelton, 2001; Peshkin & Shelton, 2002). An example of the gradient estimation via G(PO)MDP (Baxter & Bartlett, 2001) with importance sampling is given by

$$\nabla_\theta J_\pi = \mathbb{E}\left[\sum_{t=0}^{T-1} \rho_t \left(\prod_{j=0}^{t-1} \gamma_j\right) r_t \sum_{i=0}^{t} \nabla_\theta \log \pi_\theta(\mathbf{a}_i|\mathbf{s}_i)\right], \tag{5.3}$$

where $\rho_t = \prod_{z=0}^{t} \pi_\theta(\mathbf{a}_z|\mathbf{s}_z)/\pi_\beta(\mathbf{a}_z|\mathbf{s}_z)$. This technique applies only to stochastic policies and requires the knowledge of the behavioral policy π_β. Moreover, Equation (5.3) shows that path-wise importance sampling (PWIS) needs a trajectory-based dataset, since it needs to keep track of the past in the correction term ρ_t, hence introducing more restrictions on its applicability. Additionally, importance sampling suffers from high variance (Owen, 2013). Recent works have helped to make PWIS more reliable. For example, (Imani et al., 2018), building on the emphatic weighting framework (Sutton et al., 2016), proposed a trade-off between PWIS and semi-gradient approaches. Another possibility consists in restricting the gradient improvement to a safe-region, where the importance sampling does not suffer from too high variance (Metelli et al., 2018). Another interesting line of research is to estimate the importance sampling correction on a state-distribution level instead of on the classic trajectory level (Liu et al., 2018, 2019; Nachum et al., 2019). We note that, all these promising algorithms have been applied on low-dimensional problems, as importance sampling suffers from the curse of dimensionality. Our proposed solution suffers also from this problem, but we believe that our approach, as well as these recent advances in importance sampling correction, might serve for further well-theoretically-defined off-policy techniques.

Model Based

Another natural approach which comes to mind when thinking about off-policy optimization, is to use a learned model of the transition. This model allows to generate new samples and therefore to optimize the policy potentially off-line. The proclaimed efficiency of model-based techniques relies on the fact that they allow off-policy optimization. However, model-based techniques are also problematic: the model error propagates in the Bellman recursion (or in the number of steps, if we prefer), often resulting in bad policy improvements. PILCO (Deisenroth & Rasmussen, 2011) aims to optimize the policy using probabilistic inference based on Gaussian Processes to model the estimation's uncertainty. However, it works on a finite horizon setting, is restricted to unimodal state-transitions, and a particular shape of reward. PETS (Chua et al., 2018), an improved version of PILCO, builds a probabilitstic model using a bootstrapped ensemble of neural-networks, and propagates the state-distribution using particles. This method, still requires a finite horizon. Furthermore, PETS does not make use of a parametrized policy, but instead a model predictive control. The controller requires multiple neural network evaluations, which can result in an issue when interacting with a real-time system. Our method, in contrast, works on the infinite-horizon setting, and the usage of a parametrized policy is more suitable for real-time operations.

5.2. Nonparametric Off-Policy Policy Gradient

In this section we introduce a nonparametric Bellman equation with a closed form solution, which carries the dependency from the policy's parameters. We derive the gradient of the solution, and discuss the properties of the proposed estimator.

5.2.1. A Nonparametric Bellman Equation

Nonparametric Bellman equations have been developed in a number of prior works. (Ormoneit & Sen, 2002; Xu et al., 2007; Engel et al., 2005) used nonparametric models such as Gaussian Processes for approximate dynamic programming. (Taylor & Parr, 2009) have shown that these methods differ mainly in their use of regularization. (Kroemer & Peters, 2011) provided a Bellman equation using kernel density-estimation and a general overview on nonparametric dynamic programming. In contrast to prior work, our formulation preserves the dependency on the policy, enabling the computation of the policy gradient in closed-form. Moreover, we upper-bound the bias of the Nadaraya-Watson kernel regression to prove that our value function estimate is consistent w.r.t. the classical Bellman equation under smoothness assumptions. We focus on the maximization of the average return in the infinite horizon case formulated as a starting state objective $\int_{\mathbf{s}} \mu_0(\mathbf{s}) V_\pi(\mathbf{s}) \, d\mathbf{s}$ (Sutton et al., 2000).

Definition 1. *The discounted infinite-horizon objective is defined by $J_\pi = \int \mu_0(\mathbf{s}) V_\pi(\mathbf{s}) \, d\mathbf{s}$. Under a stochastic policy the objective is subject to the Bellman equation constraint*

$$V_\pi(\mathbf{s}) = \int_{\mathcal{A}} \pi_\theta(\mathbf{a}|\mathbf{s}) \left(r(\mathbf{s}, \mathbf{a}) + \gamma \int_{\mathcal{S}} V_\pi(\mathbf{s}') p(\mathbf{s}'|\mathbf{s}, \mathbf{a}) \, d\mathbf{s}' \right) d\mathbf{a}, \tag{5.4}$$

while in the case of a deterministic policy the constraint is given as

$$V_\pi(\mathbf{s}) = r(\mathbf{s}, \pi_\theta(\mathbf{s})) + \gamma \int_{\mathcal{S}} V_\pi(\mathbf{s}') p(\mathbf{s}'|\mathbf{s}, \pi_\theta(\mathbf{s})) \, d\mathbf{s}'.$$

Maximizing the objective in Definition 1 analytically is not possible, excluding special cases such as under linear-quadratic assumptions (Borrelli et al., 2017), or finite state-action space. Extracting an expression for the gradient of J_π w.r.t. the policy parameters θ is also not straightforward given the infinite set of possibly non-convex constraints represented in the recursion over V_π. Nevertheless, it is possible to transform the constraints in Definition 1 to a finite set of linear constraints via nonparametric modeling, thus leading to an expression of the value function with simple algebraic manipulation (Kroemer & Peters, 2011).

Nonparametric Modeling.

Assume a set of n observations $D \equiv \{\mathbf{s}_i, \mathbf{a}_i, r_i, \mathbf{s}'_i, \gamma_i\}_{i=1}^n$ sampled from interaction with an environment, with $\mathbf{s}_i, \mathbf{a}_i \sim \beta(\cdot, \cdot)$, $\mathbf{s}'_i \sim p(\cdot|\mathbf{s}_i, \mathbf{a}_i)$, $r_i \sim R(\mathbf{s}_i, \mathbf{a}_i)$ and $\gamma \sim \gamma(\mathbf{s}_i, \mathbf{a}_i, \mathbf{s}'_i)$. We define the kernels $\psi : \mathcal{S} \times \mathcal{S} \to \mathbb{R}^+$, $\varphi : \mathcal{A} \times \mathcal{A} \to \mathbb{R}^+$ and $\phi : \mathcal{S} \times \mathcal{S} \to \mathbb{R}^+$, as normalized, symmetric and positive definite functions with bandwidths $\mathbf{h}_\psi, \mathbf{h}_\varphi, \mathbf{h}_\phi$ respectively. We define $\psi_i(\mathbf{s}) = \psi(\mathbf{s}, \mathbf{s}_i)$, $\varphi_i(\mathbf{a}) = \varphi(\mathbf{a}, \mathbf{a}_i)$, and $\phi_i(\mathbf{s}) = \phi(\mathbf{s}, \mathbf{s}'_i)$. Following (Kroemer

& Peters, 2011), the mean reward $r(\mathbf{s}, \mathbf{a})$ and the transition conditional $p(\mathbf{s}'|\mathbf{s}, \mathbf{a})$ are approximated by the Nadaraya-Watson regression (Nadaraya, 1964; Watson, 1964) and kernel density estimation, respectively

$$\hat{r}(\mathbf{s}, \mathbf{a}) \triangleq \frac{\sum_{i=1}^n \psi_i(\mathbf{s})\varphi_i(\mathbf{a})r_i}{\sum_{i=1}^n \psi_i(\mathbf{s})\varphi_i(\mathbf{a})}, \qquad \hat{p}(\mathbf{s}'|\mathbf{s}, \mathbf{a}) \triangleq \frac{\sum_{i=1}^n \phi_i(\mathbf{s}')\psi_i(\mathbf{s})\varphi_i(\mathbf{a})}{\sum_{i=1}^n \psi_i(\mathbf{s})\varphi_i(\mathbf{a})},$$

$$\hat{\gamma}(\mathbf{s}, \mathbf{a}, \mathbf{s}') \triangleq \frac{\sum_{i=1}^n \gamma_i\psi_i(\mathbf{s})\varphi_i(\mathbf{a})\phi_i(\mathbf{s}')}{\sum_{i=1}^n \psi_i(\mathbf{s})\varphi_i(\mathbf{a})\phi_i(\mathbf{s}')}$$

Therefore, by the product of $\hat{p}$ and $\hat{\gamma}$, we obtain

$$\hat{p}_\gamma(\mathbf{s}'|\mathbf{s}, \mathbf{a}) \quad \triangleq \quad \hat{p}(\mathbf{s}'|\mathbf{s}, \mathbf{a})\hat{\gamma}(\mathbf{s}, \mathbf{a}, \mathbf{s}') = \frac{\sum_{i=1}^n \gamma_i\psi_i(\mathbf{s})\varphi_i(\mathbf{a})\phi_i(\mathbf{s}')}{\sum_{i=1}^n \psi_i(\mathbf{s})\varphi_i(\mathbf{a})}.$$

Inserting the reward and transition kernels into the Bellman Equation for the stochastic policy case, we obtain the nonparametric Bellman equation (NPBE)

$$\hat{V}_\pi(\mathbf{s}) = \int_{\mathcal{A}} \pi_\theta(\mathbf{a}|\mathbf{s})\left(\hat{r}(\mathbf{s}, \mathbf{a}) + \int_{\mathcal{S}} \hat{V}_\pi(\mathbf{s}')\hat{p}_\gamma(\mathbf{s}'|\mathbf{s}, \mathbf{a})\, \mathrm{d}\mathbf{s}'\right)\mathrm{d}\mathbf{a}$$

$$= \sum_i \int_{\mathcal{A}} \frac{\pi_\theta(\mathbf{a}|\mathbf{s})\psi_i(\mathbf{s})\varphi_i(\mathbf{a})}{\sum_j \psi_j(\mathbf{s})\varphi_j(\mathbf{a})}\, \mathrm{d}\mathbf{a}\left(r_i + \gamma_i\int_{\mathcal{S}} \phi_i(\mathbf{s}')\hat{V}_\pi(\mathbf{s}')\, \mathrm{d}\mathbf{s}'\right). \tag{5.5}$$

Equation (5.5) can be conveniently expressed in matrix form by introducing the vector of responsibilities $\varepsilon_i(\mathbf{s}) = \int \pi_\theta(\mathbf{a}|\mathbf{s})\psi_i(\mathbf{s})\varphi_i(\mathbf{a})/\sum_j \psi_j(\mathbf{s})\varphi_j(\mathbf{a})\, \mathrm{d}\mathbf{a}$, which assigns each state $\mathbf{s}$ a weight relative to its distance to a sample i under the current policy.

Definition 2. *The nonparametric Bellman equation on the dataset D is formally defined as*

$$\hat{V}_\pi(\mathbf{s}) = \varepsilon_\pi^\mathsf{T}(\mathbf{s})\left(\mathbf{r} + \int_{\mathcal{S}} \phi_\gamma(\mathbf{s}')\hat{V}_\pi(\mathbf{s}')\, \mathrm{d}\mathbf{s}'\right), \tag{5.6}$$

with $\phi_\gamma(\mathbf{s}) = [\gamma_1\phi_1(\mathbf{s}), \ldots, \gamma_n\phi_n(\mathbf{s})]^\mathsf{T}, \mathbf{r} = [r_1, \ldots, r_n]^\mathsf{T}, \varepsilon_\pi(\mathbf{s}) = [\varepsilon_1^\pi(\mathbf{s}), \ldots, \varepsilon_n^\pi(\mathbf{s})]^\mathsf{T},$

$$\varepsilon_i^\pi(\mathbf{s}) = \begin{cases} \int \pi_\theta(\mathbf{a}|\mathbf{s})\frac{\psi_i(\mathbf{s})\varphi_i(\mathbf{a})}{\sum_j \psi_j(\mathbf{s})\varphi_j(\mathbf{a})}\, \mathrm{d}\mathbf{a} & \textit{if } \pi \textit{ is stochastic} \\ \frac{\psi_i(\mathbf{s})\varphi_i(\pi_\theta(\mathbf{s}))}{\sum_j \psi_j(\mathbf{s})\varphi_j(\pi_\theta(\mathbf{s}))} & \textit{otherwise.} \end{cases}$$

From Equation (5.6) we deduce that the value function must be of the form $\varepsilon_\pi^\mathsf{T}(\mathbf{s})\mathbf{q}$, indicating that it can also be seen as a form of Nadaraya-Watson kernel regression,

$$\varepsilon_\pi^\mathsf{T}(\mathbf{s})\mathbf{q} = \varepsilon_\pi^\mathsf{T}(\mathbf{s})\left(\mathbf{r} + \int_{\mathcal{S}} \phi_\gamma(\mathbf{s}')\varepsilon_\pi^\mathsf{T}(\mathbf{s}')\mathbf{q}\, \mathrm{d}\mathbf{s}'\right). \tag{5.7}$$

Notice that, trivially, every $\mathbf{q}$ which satisfies

$$\mathbf{q} = \mathbf{r} + \int_{\mathcal{S}} \phi_\gamma(\mathbf{s}')\varepsilon_\pi^\mathsf{T}(\mathbf{s}')\mathbf{q}\, \mathrm{d}\mathbf{s}' \tag{5.8}$$

also satisfies Equation (5.7). Theorem 1 demonstrates that the algebraic solution of Equation (5.8) is the *only* solution of the nonparametric Bellman Equation (5.6).

Theorem 1. *The nonparametric Bellman equation has a unique fixed-point solution*

$$\hat{V}_\pi^*(\mathbf{s}) \stackrel{\triangle}{=} \varepsilon_\pi^\mathsf{T}(\mathbf{s})\Lambda_\pi^{-1}\mathbf{r},$$

with $\Lambda_\pi \stackrel{\triangle}{=} I - \hat{\mathbf{P}}_\pi^\gamma$ *and* $\hat{\mathbf{P}}_\pi^\gamma \stackrel{\triangle}{=} \int_S \phi_\gamma(\mathbf{s}')\varepsilon_\pi^\mathsf{T}(\mathbf{s})\,\mathrm{d}\mathbf{s}'$, *where* Λ_π *is always invertible since* $\hat{\mathbf{P}}^{\pi,\gamma}$ *is a strictly sub-stochastic matrix (Frobenius' Theorem). The statement is valid also for* $n \to \infty$, *provided bounded R.*

Proof of Theorem 1 is provided in Appendix B.

5.2.2. Nonparametric Gradient Estimation

With the closed-form solution of $\hat{V}_\pi^*(\mathbf{s})$ from Theorem 1, it is possible to compute the analytical gradient of J_π w.r.t. the policy parameters

$$\begin{aligned}
\nabla_\theta \hat{V}_\pi^*(\mathbf{s}) &= \left(\frac{\partial}{\partial\theta}\varepsilon_\pi^\mathsf{T}(\mathbf{s})\right)\Lambda_\pi^{-1}\mathbf{r} + \varepsilon_\pi^\mathsf{T}(\mathbf{s})\frac{\partial}{\partial\theta}\Lambda_\pi^{-1}\mathbf{r} \\
&= \underbrace{\left(\frac{\partial}{\partial\theta}\varepsilon_\pi^\mathsf{T}(\mathbf{s})\right)\Lambda_\pi^{-1}\mathbf{r}}_{A} + \underbrace{\varepsilon_\pi^\mathsf{T}(\mathbf{s})\Lambda_\pi^{-1}\left(\frac{\partial}{\partial\theta}\hat{\mathbf{P}}_\pi^\gamma\right)\Lambda_\pi^{-1}\mathbf{r}}_{B} .
\end{aligned} \tag{5.9}$$

Substituting the result of Equation (5.9) into the return specified in Definition 1, introducing $\varepsilon_{\pi,0}^\mathsf{T} \stackrel{\triangle}{=} \int \mu_0(\mathbf{s})\varepsilon_\pi^\mathsf{T}(\mathbf{s})\,\mathrm{d}\mathbf{s}$, $\mathbf{q}_\pi = \Lambda_\pi^{-1}\mathbf{r}$, and $\mu_\pi = \Lambda_\pi^{-\mathsf{T}}\varepsilon_{\pi,0}$ we obtain

$$\nabla_\theta \hat{J}_\pi = \left(\frac{\partial}{\partial\theta}\varepsilon_{\pi,0}^\mathsf{T}\right)\mathbf{q}_\pi + \mu_\pi^\mathsf{T}\left(\frac{\partial}{\partial\theta}\hat{\mathbf{P}}_\pi^\gamma\right)\mathbf{q}_\pi, \tag{5.10}$$

where $\mathbf{q}_\pi$ and μ_π can be estimated via conjugate gradient to avoid the inversion of Λ_π.

The terms A and B in Equation (5.9) correspond to the terms in Equation (5.2). In contrast to semi-gradient actor-critic methods, where the gradient bias is affected by both the critic bias and the semi-gradient approximation (Imani et al., 2018; Fujimoto et al., 2019), our estimate is the *full gradient* and the only source of bias is introduced by the estimation of $\hat{V}_\pi$, which we analyze in Section 5.2.3. The term μ_π can be interpreted as the support of the state-distribution as it satisfies $\mu_\pi^\mathsf{T} = \varepsilon_{\pi,0}^\mathsf{T} + \mu_\pi^\mathsf{T}\hat{\mathbf{P}}_\pi^\gamma$. In Section 5.3, more specifically in Figure 5.5, we empirically show that $\varepsilon_\pi^\mathsf{T}(\mathbf{s})\mu_\pi$ provides an estimate of the state distribution over the whole state-space. Implementation-wise, the quantities $\varepsilon_{\pi,0}^\mathsf{T}$ and $\hat{\mathbf{P}}_{i,j}^\pi$ are estimated via Monte-Carlo integration, which is unbiased but computationally demanding, or using other techniques such as unscented transform or numerical quadrature. The matrix $\hat{\mathbf{P}}_\pi^\gamma$ is of dimension $n \times n$, which can be memory-demanding. In practice, we notice that the matrix is often sparse. By taking advantage of conjugate gradient and sparsification we are able to achieve computational complexity of $\mathcal{O}(n^2)$ per policy update and memory complexity of $\mathcal{O}(n)$. A schematic of our implementation is summarized in Algorithm 2.

5.2.3. Theoretical Analysis

Nonparametric estimates of the transition dynamics and reward enjoy favorable properties for an off-policy learning setting. A well-known asymptotic behavior of the Nadaraya-Watson kernel regression,

$$\mathbb{E}\left[\lim_{n\to\infty}\hat{f}_n(x)\right] - f(x) \approx h_n^2\left(\frac{1}{2}f''(x) + \frac{f'(x)\beta'(x)}{\beta(x)}\right)\int u^2 K(u)\,\mathrm{d}u,$$

Algorithm 2 Nonparametric Off-Policy Policy Gradient

1: **input:** dataset $\{\mathbf{s}_i, \mathbf{a}_i, r_i, \mathbf{s}'_i, \gamma_i\}_{i=1}^{n}$ where π_θ indicates the policy to optimize and ψ, ϕ, φ the kernels respectively for state, action and next-state.
2: **while** not converged **do**
3: $\quad$ Compute $\boldsymbol{\varepsilon}_\pi^{\mathsf{T}}(\mathbf{s})$ as in Definition 2 and $\boldsymbol{\varepsilon}_{\pi,0}^{\mathsf{T}} \overset{\triangle}{=} \int \mu_0(\mathbf{s})\boldsymbol{\varepsilon}_\pi^{\mathsf{T}}(\mathbf{s})\,\mathrm{d}\mathbf{s}$.
4: $\quad$ Estimate $\hat{\mathbf{P}}_\pi^\gamma$ as defined in Theorem 1 using MC integration ($\phi(\mathbf{s})$ is a distribution).
5: $\quad$ Solve $\mathbf{r} = \boldsymbol{\Lambda}_\pi \mathbf{q}_\pi$ and $\boldsymbol{\varepsilon}_{\pi,0} = \boldsymbol{\Lambda}_\pi^{\mathsf{T}} \boldsymbol{\mu}_\pi$ for $\mathbf{q}_\pi$ and $\boldsymbol{\mu}_\pi$ using conjugate gradient.
6: $\quad$ Update θ using Equation (5.10).
7: **end while**

---

shows how the bias is related to the regression function $f(x)$, as well as to the samples' distribution $\beta(x)$ (Fan, 1992; Wasserman, 2006). However, this asymptotic behavior is valid only for infinitesimal bandwidth, infinite samples ($h \to 0, nh \to \infty$) and requires the knowledge of the regression function and of the sampling distribution.

In a recent work, we propose an upper bound of the bias that is also valid for finite bandwidths (Tosatto et al., 2020a). We show under some Lipschitz conditions that the bound of the Nadaraya-Watson kernel regression bias does not depend on the samples' distribution, which is a desirable property in off-policy scenarios. The analysis is extended to multidimensional input space. For clarity of exposition, we report the main result in its simplest formulation, and later use it to infer the bound of the NPBE bias.

Theorem 2. *Let $f : \mathbb{R}^d \to \mathbb{R}$ be a Lipschitz continuous function with constant L_f. Assume a set $\{\mathbf{x}_i, y_i\}_{i=1}^{n}$ of i.i.d. samples from a log-Lipschitz distribution β with a Lipschitz constant L_β. Assume $y_i = f(\mathbf{x}_i) + \epsilon_i$, where $f : \mathbb{R}^d \to \mathbb{R}$ and ϵ_i is i.i.d. and zero-mean. The bias of the Nadaraya-Watson kernel regression with Gaussian kernels in the limit of infinite samples $n \to \infty$ is bounded by*

$$\left| \mathbb{E}\left[\lim_{n \to \infty} \hat{f}_n(\mathbf{x}) \right] - f(\mathbf{x}) \right| \leq \frac{L_f \sum\limits_{k=1}^{d} \mathbf{h}_k \left(\prod\limits_{i \neq k}^{d} \chi_i \right) \left(\frac{1}{\sqrt{2\pi}} + \frac{L_\beta \mathbf{h}_k}{2} \chi_k \right)}{\prod\limits_{i=1}^{d} e^{\frac{L_\beta^2 \mathbf{h}_i^2}{2}} \left(1 - \mathrm{erf}\left(\frac{\mathbf{h}_i L_\beta}{\sqrt{2}} \right) \right)},$$

where

$$\chi_i = e^{\frac{L_\beta^2 \mathbf{h}_i^2}{2}} \left(1 + \mathrm{erf}\left(\frac{\mathbf{h}_i L_\beta}{\sqrt{2}} \right) \right),$$

$\mathbf{h} > 0 \in R^d$ *is the vector of bandwidths and* erf *is the error function.*

Building on Theorem 2 we show that the solution of the NPBE is consistent with the solution of the true Bellman equation. Moreover, although the bound is not affected directly by $\beta(\mathbf{s})$, a smoother sample distribution $\beta(\mathbf{s})$ plays favorably in the bias term (a low L_β is preferred).

Theorem 3. *Consider an arbitrary MDP $\mathcal{M}$ with a transition density p and a stochastic reward function $R(\mathbf{s}, \mathbf{a}) = r(\mathbf{s}, \mathbf{a}) + \epsilon_{\mathbf{s}, \mathbf{a}}$ where $r(\mathbf{s}, \mathbf{a})$ is a Lipschitz continuous function with L_R constant and $\epsilon_{\mathbf{s}, \mathbf{a}}$ denotes zero-mean noise. Assume $|R(\mathbf{s}, \mathbf{a})| \leq R_{max}$ and a dataset D_n sampled from a log-Lipschitz distribution β defined over the state-action space with Lipschitz constant L_β. Let V_D be the unique solution of a nonparametric Bellman* *equation with Gaussian kernels ψ, φ, ϕ with positive bandwidths $\mathbf{h}_\psi, \mathbf{h}_\varphi, \mathbf{h}_\phi$ defined over the dataset $\lim_{n\to\infty} D_n$. Assume V_D to be Lipschitz continuous with constant L_V. The bias of such estimator is bounded by*

$$\left|\overline{V}(\mathbf{s}) - V^\pi(\mathbf{s})\right| \leq \frac{1}{1-\gamma_c}\left(A_{Bias} + \gamma_c L_V \sum_{k=1}^{d_s} \frac{h_{\phi,k}}{\sqrt{2\pi}}\right), \tag{5.11}$$

where $\overline{V}(\mathbf{s}) = \mathbb{E}_D[V_D(\mathbf{s})]$, A_{Bias} is the bound of the bias provided in Theorem 2 with $L_f = L_R$, $\mathbf{h} = [\mathbf{h}_\psi, \mathbf{h}_\varphi]$, $d = d_s + d_a$ and $V^\pi(\mathbf{s})$ is the fixed point of the ordinary Bellman equation. [2]

Theorem 3 shows that the value function provided by Theorem 1 is consistent when the bandwidth approaches infinitesimal values. Moreover, it is interesting to notice that the error can be decomposed in A_{Bias}, which is the bias component dependent on the reward's approximation, and the remaining term that depends on the smoothness of the value function and the bandwidth of ϕ, which can be read as the error of the transition's model. For detailed proof of Theorem 3 refer to Appendix B.

The independence from the sampling distribution suggests that, under these assumptions, nonparametric estimation is particularly suited for off-policy setting, as the bias is not affected by different behavioral policies. More in detail, the bound shows that smoother reward functions, state-transitions and sample distributions play favorably against the estimation bias.

A Trust Region

Very commonly, in order to prevent harmful policy optimization, the policy is constrained to stay close to the data (Peters et al., 2010), to avoid taking large steps (Schulman et al., 2015, 2017) or to circumvent large variance in the estimation (Metelli et al., 2018; Chua et al., 2018). These techniques, which keep the policy updates in a trusted-region, prevent incorrect and dangerous estimates of the gradient. Even if we do not include any explicit constraint of this kind, the Nadaraya-Watson kernel regression automatically discourages policy improvements towards low-data areas. In fact, as depicted in Figure 5.2, the Nadaraya-Watson kernel regression, tends to predict a constant function in low density regions. Usually, this characteristic is regarded as an issue, as it causes the so-called boundary-bias. In our case, this effect turns out to be beneficial, as it constrains the policy to stay close to the samples, where the model is more correct.

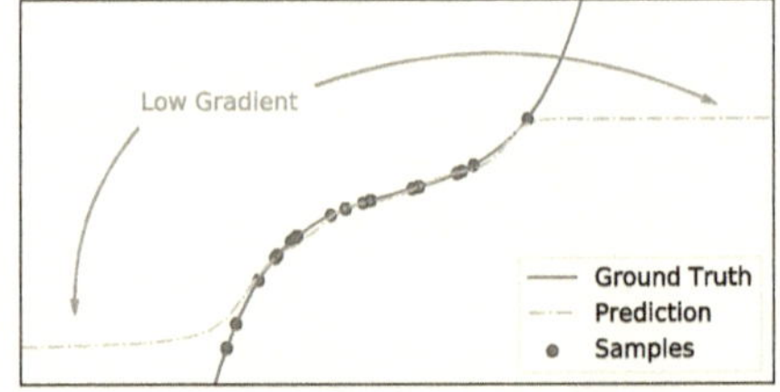

Figure 5.2.: The classic effect (known as boundary-bias) of the Nadaraya-Watson regression predicting a constant function in low-density regions is beneficial in our case, as it prevents the policy from moving in those areas as the gradient gets close to zero.

5.3. Empirical Analysis

In this section, we analyze our method. Therefore, we divide our experiments in two logical sections: the analysis of the gradient, and the analysis of the policy optimization using a gradient ascent technique. The

---

[2] Complete proofs of the theorems and precise definitions can be found in the supplementary material.

analysis of the gradient comprises an empirical evaluation of the bias, the variance and the gradient direction w.r.t. the ground truth, in relation to some quantities such as the size of the dataset or its degree of "off-policiness". In the policy optimization analysis, instead, we aim to both compare the sample efficiency of our method in comparison to state-of-the-art policy gradient algorithms, and to study its applicability to unstructured and human-demonstrated datasets.

5.3.1. Benchmarking Tasks

In the following, we give a brief description of the tasks involved in the empirical analysis.

Linear Quadratic Gaussian Controller

A very classical control problem consists of linear dynamics, quadratic reward and Gaussian noise. The main advantage of this control problem relies in the fact that it is fully solvable in closed-form, using the Riccati equations, which makes it appropriate for verifying the correctness of our algorithm. In our specific scenario, we have a policy encoded with two parameters for illustration purposes. The LQG is defined as

$$\max_{\boldsymbol{\theta}} \mathbb{E}\left[\sum_{t=0}^{\infty} \gamma^t r_t\right]$$
$$\text{s.t.} \quad \mathbf{s}_{t+1} = A\mathbf{s}_t + B\mathbf{a}_t,$$
$$r_t = -\mathbf{s}_t^{\mathsf{T}} Q \mathbf{s}_t - \mathbf{a}_t^{\mathsf{T}} R \mathbf{a}_t,$$
$$\mathbf{a}_{t+1} = \Theta \mathbf{s}_t + \Sigma \epsilon_t,$$
$$\epsilon_t \sim \mathcal{N}(0, I),$$

with A, B, Q, R, Σ diagonal matrix and $\Theta = \mathrm{diag}(\boldsymbol{\theta})$ where $\boldsymbol{\theta}$ are considered the policy's parameters. In the stochastic policy experiments, $\pi_{\theta}(\mathbf{a}|\mathbf{s}) = \mathcal{N}(\mathbf{a}|\Theta \mathbf{s}; \Sigma)$, while for the deterministic case $\Sigma = \mathbf{0}$ and $\pi_{\theta}(\mathbf{s}) = \Theta \mathbf{s}$. For further details, please refer to Appendix B.

OpenAI Pendulum-v0

The OpenAI Pendulum-v0 (Brockman et al., 2016) is a popular benchmark in reinforcement learning. It simulates a simple under-actuated inverted-pendulum. The goal is to swing the pendulum until it reaches the top position, and then to keep it stable. The state of the system is fully described by the angle of the pendulum ω and its angular velocity $\dot{\omega}$. The applied torque $\tau \in [-2, 2]$ corresponds to the agent's action. One of the advantages of such a system, is that its well-known value function is two-dimensional.

Quanser Cart-pole

The cart-pole is another classical task in reinforcement learning. It consists of an actuated cart moving on a track, to which a pole is attached. The goal is to actuate the cart in a way to balance the pole in the top position. Differently from the inverted pendulum, the system has a further degree of complexity, and the state space requires the position on the track x, the velocity of the cart $\dot{x}$, the angle of the pendulum ω and its angular velocity $\dot{\omega}$.

Acronym	Description	Typology
NOPG-D	Our method with deterministic policy.	NOPG
NOPG-S	Our method with stochastic policy.	
G(PO)MDP+N	G(PO)MDP with normalized importance sampling.	PWIS
G(PO)MDP+BN	G(PO)MDP with normalized importance sampling and generalized baselines.	
DPG+Q	Offline version of the deterministic policy gradient theorem with an oracle for the Q-function.	SG
DDPG-Off	Offline version of the deep deterministic policy gradient theorem.	
DDPG-On	Classic version of DDPG.	
TD3	Improved version of DDPG.	
SAC	Classic version of SAC.	

Table 5.1.: Acronyms used in the chapter to refer to practical implementation of the algorithms.

OpenAI Mountain-Car

The mountain-car (also known as car-on-hill), consists on an under-powered car that must reach the top of a hill. The car is placed in the valley connecting two hills. In order to reach the goal position, it must first go in opposite direction in order to gain momentum. Its state is described by the x-position of the car, and by its velocity $\dot{x}$. The episodes terminate when the car reaches the goal. In contrast to the swing-up pendulum, which is hardly controllable by a human-being, this car system is ideal to provide human-demonstrated data.

5.3.2. Algorithms Used for Comparisons

In order to provide an analysis of the gradient, we compare our algorithm against G(PO)MDP with importance sampling, and with off-line DPG. Instead of using the naïve form of G(PO)MDP with importance sampling, which suffers from high variance, we used the normalized importance sampling (Shelton, 2013; Rubinstein & Kroese, 2016) (which introduces some bias but drastically reduces the variance), and the generalized baselines (Jie & Abbeel, 2010) (which also introduce some bias, as they are estimated from the same dataset). The off-line version of DPG, suffers from three different sources of bias: the semi-gradient, the critic approximation and the improper use of the discounted state distribution (Thomas, 2014; Nota & Thomas, 2020). To focus on the semi-gradient contribution to the bias, we provide an oracle Q-function that does not suffer of critic approximation and improper discounting (we denote this version as DPG+Q). For the policy improvement, instead, we compare to more sophisticated and recent deep reinforcement learning techniques, such as TD3 (Fujimoto et al., 2019) and SAC (Haarnoja et al., 2018). A full list of the algorithms used in the comparisons with a brief description is available in Table 5.1.

5.3.3. Analysis of the Gradient

We want to compare the bias and variance of our gradient estimator w.r.t. the already discussed classical estimators. Therefore, we use the LQG setting described in Section 5.3.1, which allows us to compute the true gradient. Our goal is to estimate the gradient w.r.t. the policy π_θ diagonal parameters θ_1, θ_2, while sampling from a policy which is a linear combination of Θ and Θ'. The hyper-parameter α determines the mixing

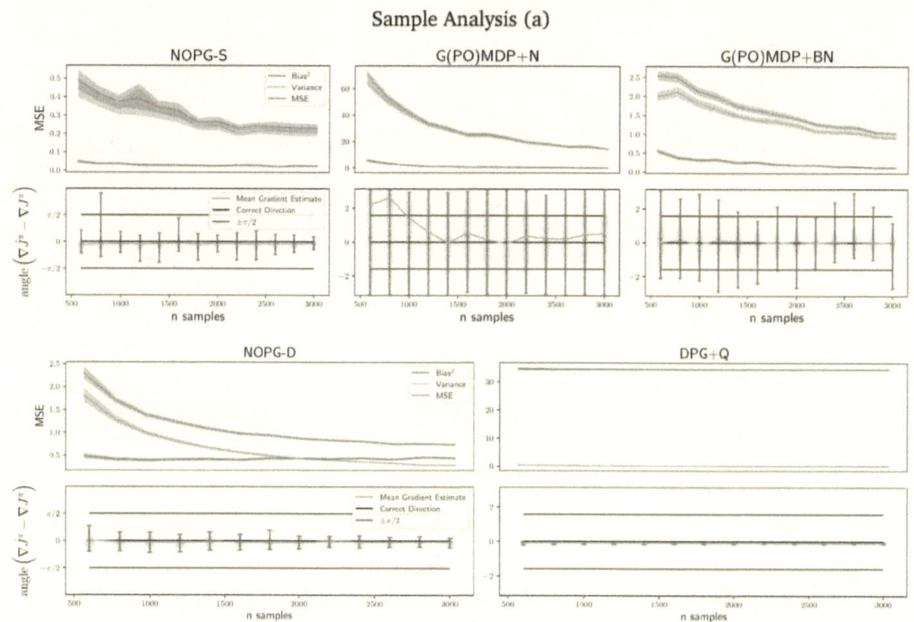

Figure 5.3.: Bias, variance, MSE and gradient direction with respect to the number of samples. Our method enjoys lower mean squared error. In our analysis, DPG+Q exhibits a good direction of the gradient estimate, but his mean squared error does not decrease, suggesting (in agreement with the theory) that its estimate is not consistent.

between the two parameters. When $\alpha = 1$ the behavioral policy will have parameters Θ', while when $\alpha = 0$ the dataset will be sampled using Θ. In Figure 5.6, we can visualize the difference of the two policies with parameters Θ and Θ'. Although not completely disjoint, they are fairly far in the probability space, especially if we take into account that such distance propagates in the length of the trajectories.

Sample Analysis

We want to study how the bias, the variance and the direction of the estimated gradient vary w.r.t. the dataset's size. We are particularly interested in the off-policy strategy for sampling, and in this set of experiments we will use constant $\alpha = 0.5$. Figure 5.3 depicts these quantities w.r.t. the number of collected samples. As expected, a general trend for all algorithms is that with a higher number of samples we are able to reduce the variance. The importance sampling based G(PO)MDP algorithms eventually obtain a low bias as well. Remarkably, NOPG has significantly both lower bias and variance, and its gradient direction is also more accurate w.r.t. the G(PO)MDP algorithms (note the different scales of the y-axis). Between DPG+Q and NOPG there is no sensible difference, but we should take into account the already-mentioned advantage of DPG+Q to have access to the true Q-function.

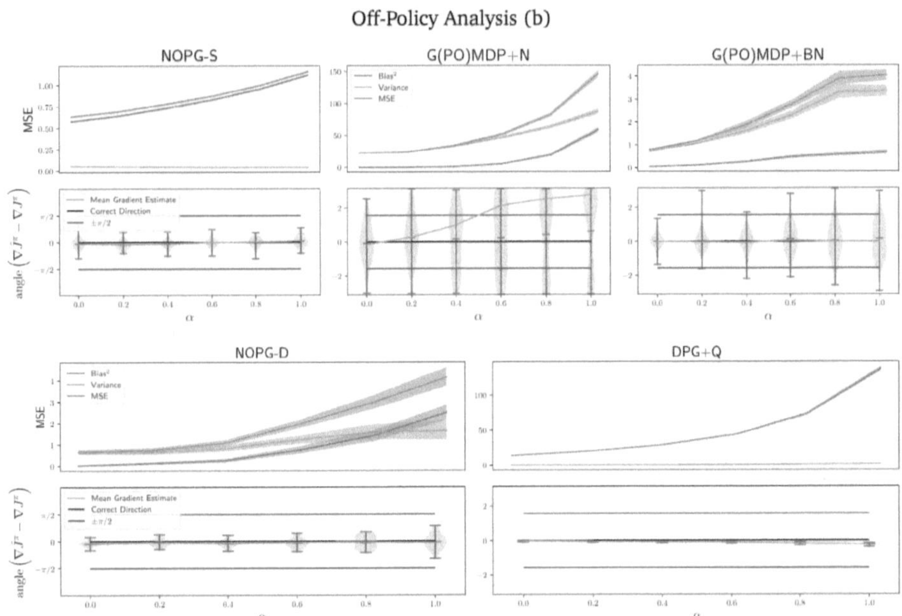

Figure 5.4.: Bias, variance, MSE and gradient direction w.r.t. the "off-policiness" of the samples. Our algorithm exhibits a lower mean squared error and a more consistent gradient direction when compared to the baselines.

Off-Policy Analysis

We want to estimate the bias and the variance w.r.t. different degrees of "off-policiness" α. We want to highlight that in the deterministic experiment the behavioral policy remains stochastic. This is needed to ensure the stochastic generation of datasets, which is essential to estimate the bias and the variance of the estimator. As depicted in Figure 5.4, the variance in importance sampling based techniques tends to increase when the dataset is off-policy. On the contrary, NOPG seems to be more subject to an increase of bias. This trend is also noticeable in DPG+Q, where the component of the bias is the one playing a major role in the mean squared error. The gradient direction of NOPG seems however unbiased, while DPG+Q has a slight bias but remarkably less variance (note the different scales of the y-axis). We remark that DPG+Q uses an oracle for the Q-function, which supposedly results in lower variance and bias[3]. The positive bias of DPG+Q in the

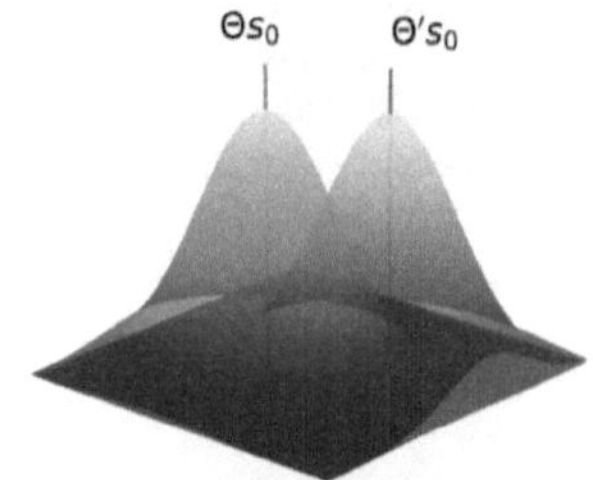

Figure 5.6.: The optimization policy having parameters θ_1, θ_2 and the behavioral policy having parameters θ_1', θ_2' exhibit a fair distance in probability space.

---

[3]Furthermore, we suspect that the particular choice of a LQG task tends to mitigate the problems of DPG, as the fast convergence to a stationary distribution due to the stable attractor, united with the improper discounting, results in a coincidental correction of

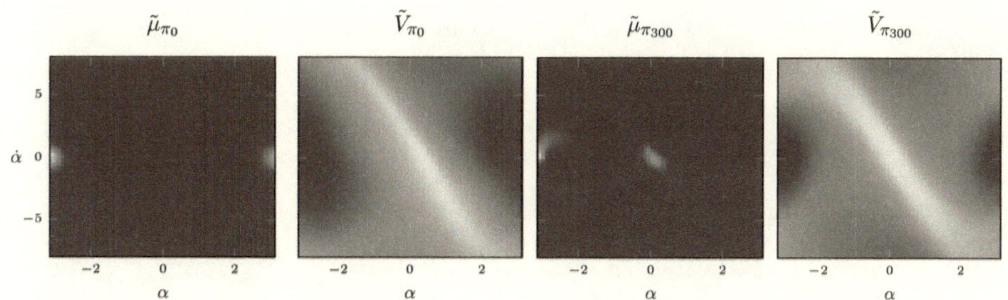

Figure 5.5.: A phase portrait of the state distribution $\tilde{\mu}_\pi$ and value function $\tilde{V}_\pi$ estimated in the swing-up pendulum task with NOPG-D. Green corresponds to higher values. The two leftmost figures show the estimates before any policy improvement, while the two rightmost show them after 300 offline updates of NOPG-D. Notice that the algorithm finds a very good approximation of the optimal value function and is able to predict that the system will reach the goal state $((\alpha, \dot{\alpha}) = (0,0))$.

on-policy case ($\alpha = 0$) is caused by the improper use of discounting. In general, NOPG shows a decrease in bias and variance in order of magnitudes when compared to the other algorithms.

Bandwidth Analysis

In the previous analysis, we kept the bandwidth's parameters of our algorithm fixed, even though a dynamic adaptation of this parameter w.r.t. the size of the dataset might have improved the bias/variance trade-off. We are now interested in studying how the bandwidth impacts the gradient estimation. For this purpose, we generated datasets of 1000 samples with $\alpha = 0.5$. We set all the bandwidths of state, action and next state, for each dimension equal to κ. From Figure 5.8 we evince that a lower bandwidth corresponds to a higher variance, while a larger bandwidth approaches a constant bias and the variance tends to zero. This result is in line with the theory.

5.3.4. Policy Improvement

In the previous section, we analyzed the statistical properties of our estimator. Conversely, in this section, we use the NOPG estimate to fully optimize the policy. At the current state, NOPG is a batch algorithm, meaning that it receives as input a set of data, and it outputs an optimized policy, without any interaction with the environment. We study the sample efficiency of the overall algorithm. We compare it with both other batch and online algorithms. Please notice that online algorithms, such as DDPG-On, TD3 and SAC, can acquire more valuable samples during the optimization process. Therefore, in a direct comparison, batch algorithms are in disadvantage.

Uniform Grid. In this experiment we analyze the performance of NOPG under a uniformly sampled dataset, since, as the theory suggests, this scenario should yield the least biased estimate of NOPG. We generate datasets from a grid over the state-action space of the pendulum environment with different granularities. We

---

the state-distribution.

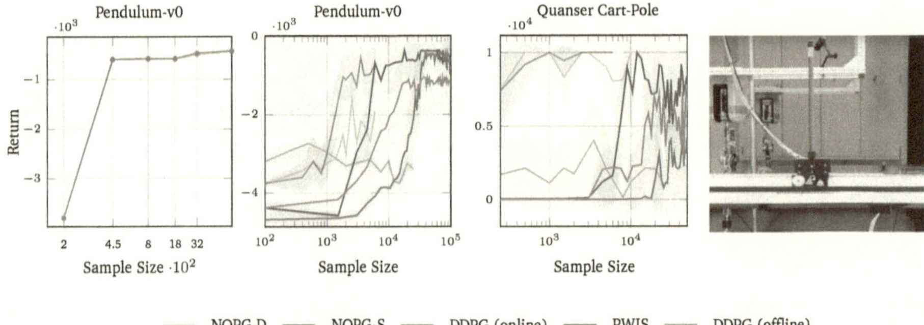

Figure 5.7.: Comparison of NOPG in its deterministic and stochastic versions to state-of-the-art algorithms on continuous control tasks: Swing-Up Pendulum with **uniform grid** sampling (left), Swing-Up Pendulum with the **random agent** (center-left) and the Cart-Pole stabilization (center-right). The figures depict the mean and 95% confidence interval over 10 trials. NOPG outperforms the baselines w.r.t the sample complexity. **Note the log-scale along the x-axis**. The right most picture shows the real cart-pole platform from Quanser.

test our algorithm by optimizing a policy encoded with a neural-network for a fixed amount of iterations. The policy is composed of a single hidden layer with 50 neurons and ReLU activations. This configuration is fixed across all the different experiments and algorithms for the remainder of this document. The resulting policy is evaluated on trajectories of 500 steps starting from the bottom position. The leftmost plot in Figure 5.7, depicts the performance against different dataset sizes, showing that NOPG is able to solve the task with 450 samples. Figure 5.5 is an example of the value function and state distribution estimates of NOPG-D at the beginning and after 300 optimization steps. The ability to predict the state-distribution is particularly interesting for robotics, as it is possible to predict in advance whether the policy will move towards dangerous states. Note that this experiment is not applicable to PWIS, as it does not admit non-trajectory-based data.

Online Setting. In contrast to the uniform grid experiment, here we collect the datasets using trajectories from a random agent in the pendulum and the cart-pole environments. In the pendulum task, the trajectories are generated starting from the up-right position and applying a policy composed of a mixture of two Gaussians. The policies are evaluated starting from the bottom position with an episode length of 500 steps. The datasets used in the cart-pole experiments are collected using a uniform policy starting from the upright position until the end of the episode, which occurs when the absolute value of the angle θ surpasses $3\,\mathrm{deg}$. The optimization policy is evaluated for 10^4 steps. The reward is $r_t = \cos\theta_t$. Since θ is defined as 0 in the top-right position, a return of 10^4 indicates an optimal policy behavior.

We analyze the sample efficiency by testing NOPG and DDPG-Off in an offline fashion with pre-collected samples, on a different number of trajectories. In addition, we provide the learning curve of DDPG-On, TD3 and SAC using the implementation in

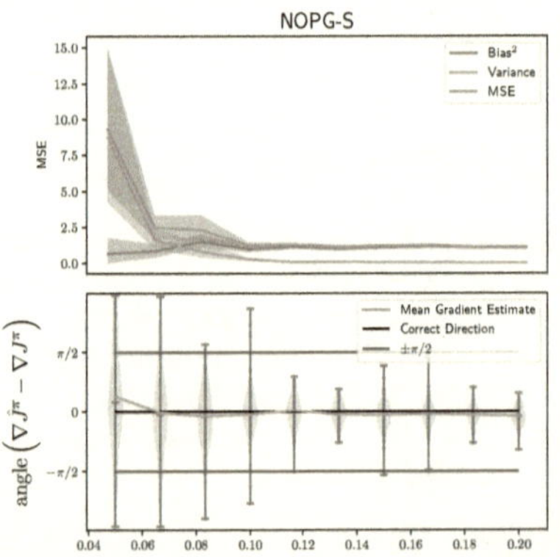

Figure 5.8.: A lower bandwidth corresponds to higher variance, while higher bandwidth increases the bias up to a plateau.

Mushroom (D'Eramo et al., 2020). For a fixed size of the dataset,
we optimize DDPG-Off and NOPG for a fixed number of steps. Since DDPG-Off exhibits an unstable learning, we select the best policy obtained during the learning process. For NOPG, instead, we select the policy from the last optimization step. The two rightmost plots in Figure 5.7 highlight that our algorithm has superior sample efficiency by more than one order of magnitude (note the log-scale on the x-axis).

To validate the resulting policy learned in simulation, we apply the final learned controller on a real Quanser cart-pole, and observe a successful stabilizing behavior as can be seen in the supplementary video.

Human Demonstrated Data. In robotics, learning from human demonstrations is crucial in order to obtain better sample efficiency and to avoid dangerous policies. This experiment is designed to showcase the ability of our algorithm to deal with such demonstrations without the need for explicit knowledge of the underlying behavioral policy. The experiment is executed in a completely offline fashion after collecting the human dataset, i.e., without any further interaction with the environment. This setting is different from the classical imitation learning and subsequent optimization (Kober & Peters, 2009). As an environment we choose the continuous mountain car task from OpenAI. We provide 10 demonstrations recorded by a human operator and assigned a reward of -1 to every step. A demonstration ends when the human operator surpasses the limit of 500 steps, or arrives at the goal position. The human operator explicitly provides sub-optimal trajectories, as we are interested in analyzing whether NOPG is able to take advantage of the human demonstrations

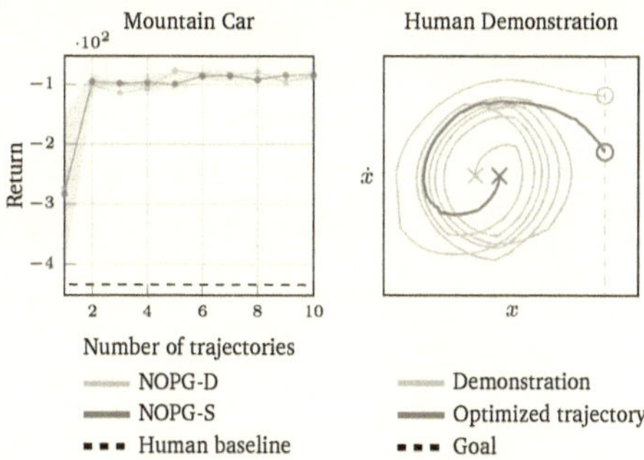

Figure 5.9.: With a small amount of data NOPG is able to reach a policy that surpasses the human demonstrator (dashed line) in the mountain car environment. Depicted are the mean and 95% confidence over 10 trials (left). An example of a human-demonstrated trajectory and the relative optimized version obtained with NOPG (right). Although the human trajectories in the dataset are sub-optimal, NOPG converges to an optimal solution (right).

to learn a better policy than that of the human, without any further interaction with the environment. To obtain a sample analysis, we evaluate NOPG on randomly selected sub-sets of the trajectories from the human demonstrations. Figure 5.9 shows the average performance as a function of the number of demonstrations as well as an example of a human-demonstrated trajectory. Notice that both NOPG-S and NOPG-D manage to learn a policy that surpasses the human operator's performance and reach the optimal policy with two demonstrated trajectories.

5.4. Epilogue

In this chapter, we presented and analyzed an off-policy gradient technique *Nonparametric Off-policy Policy Gradient* (NOPG) (Tosatto et al., 2020b). Our estimator overcomes the main issues of the techniques of off-policy gradient estimation. On the one hand, in contrast to semi-gradient approaches, it delivers a full-gradient estimate; and on the other hand, it avoids the high variance of importance sampling by phrasing the problem with nonparametric techniques. The empirical analysis clearly showed a better gradient estimate in terms of bias, variance, and direction. Our experiments also showed that our method has high sample efficiency and that our algorithm can be behavioral-agnostic and cope with unstructured data.

However, our algorithm, which is built on nonparametric techniques, suffers from the curse of dimensionality. Furthermore, it currently does not account for the exploration problem, which is important to avoid local optima. As a future work, we will study a parametric approximation of the Bellman equation, which similarly to NOPG allows for a full-gradient estimate, scales better with the number of samples. The exploration can be achieved by combining an online definition of the algorithm with classic entropy regularization techniques.

6. An Upper Bound of the Bias of Nadaraya-Watson Kernel Regression

In the previous chapter, we constructed a closed-form solution of the Bellman equation based on Nadaraya-Watson kernel regression, and we provided an upper bound on the bias estimation. In this chapter, we investigate more in-depth detail on the bias of Nadaraya-Watson kernel regression. We provide a hard (non-probabilistic) bound on a broad set of design and regression functions. We believe that similar guarantees are essential to developing safe machine learning.

6.1. Prologue

Nonparametric regression and density estimation have been used in a wide spectrum of applications, ranging from economics (Bansal et al., 1995), system dynamics identification (Wang et al., 2006; Nguyen-Tuong & Peters, 2010), and reinforcement learning (Ormoneit & Sen, 2002; Kroemer & Peters, 2011; Deisenroth & Rasmussen, 2011; Kroemer et al., 2012). In recent years, nonparametric density estimation and regression have been dominated by parametric methods such as those based on deep neural networks. These parametric methods have demonstrated an extraordinary capacity in dealing with both high-dimensional data—such as images, sounds, or videos—and large dataset. However, it is difficult to obtain strong guarantees on such complex models, which have been shown easy to fool (Moosavi-Dezfooli et al., 2016). Nonparametric techniques have the advantage of being easier to understand, and recent work overcame some of their limitations by, e.g., allowing linear-memory and sub-linear query time for density kernel estimation (Charikar & Siminelakis, 2017; Backurs et al., 2019). These methods allowed nonparametric kernel density estimation to be performed on datasets of 10^6 samples and up to 784 input dimension. As such, nonparametric methods are a suitable choice when one is willing to trade performance for statistical guarantees; and the contribution of this paper is to advance the state-of-the-art on such guarantees.

Studying the error of a statistical estimator is important. It can be used, for example, to tune the hyper-parameters by minimizing the estimated error (Härdle & Marron, 1985; Ray & Tsay, 1997; Herrmann et al., 1992; Köhler et al., 2014). To this end, the estimation error is usually decomposed into an estimation *bias* and *variance*. When it is not possible to derive these quantities, one performs an asymptotic behavior analysis or a convergence to a probabilistic distribution of the error. While all aforementioned analyses give interesting insights on the error and allow for hyper-parameter optimization, they do not provide any strong guarantee on the error, i.e., we cannot *upper bound* it with absolute certainty.

Beyond hyper-parameter optimization, we argue that another critical aspect of error analysis is to provide hard (non-probabilistic) bounds of the error for critical data-driven algorithms. We believe that learning agents taking autonomous, data-driven, decisions will be increasingly present in the near future. These agents will, for example, be autonomous surgeons, self-driving cars or autonomous manipulators. In many critical applications involving these agents, it is of primary importance to bound the prediction error in order to provide some technical guarantees on the agent's behavior. In this paper we derive a hard upper bound of the estimation bias in non-parametric regression with minimal assumptions on the problem. The bound can be readily applied to a wide range of applications.

Specifically, we consider the Nadaraya–Watson kernel regression (Nadaraya, 1964; Watson, 1964), which can be seen as a conditional kernel density estimate. We derive an upper bound of the estimation bias under weak local Lipschitz assumptions. The reason for our choice of estimator falls in its inherent simplicity compared to more sophisticated techniques. The bias of the Nadaraya–Watson kernel regression has been previously studied by Rosenblatt (1969), and has been reported in a number of related work (Mack & Müller, 1988; Fan, 1992; Fan & Gijbels, 1992; Wasserman, 2006). The analysis of the bias conducted by Rosenblatt (1969) Rosenblatt (1969) still remains the main reference for this regression technique. The main assumptions of Rosenblatt's analysis are $h_n \to 0$ (where n is the number of samples) and $nh_n \to \infty$ where h_n is the kernel's bandwidth. Rosenblatt's analysis suffers from an asymptotic error $o(h_n^2)$, which means that for large bandwidths it is not accurate; making it inapplicable to derive a hard upper bound. To the best of our knowledge, the only proposed bound on the bias requires the restrictive assumption that the samples must be placed evenly on a closed interval (Györfi et al., 2006). In contrast, we derive an upper bound of the bias of the Nadaraya–Watson kernel regression that is valid for a large class of design and for any choice of bandwidth.

We build our analysis on weak Lipschitz assumptions (Miculescu, 2000), which are milder than the (global) Lipschitz, as we require only $|f(x) - f(y)| \leq L|x - y| \; \forall y \in \mathcal{C}$ given a fixed x, instead of the classic $|f(x) - f(y)| \leq L|x - y| \; \forall y, x \in \mathcal{C}$—where $\mathcal{C}$ is the data domain. Lipschitz assumptions are common in different fields, and usually allow a wide family of admissible functions. This is particularly true when the Lipschitz is require for only a subset of the function's domain (like in our case). Moreover, notice that the classical analysis requires the knowledge of the second derivative of the regression function m, and therefore the continuity of m'. Our Lipschitz assumption is less restrictive, allowing us to obtain a bias upper bound even for functions like $m(x) = |x|$, at points (like $x = 0$) where m'' is undefined. The Rosenblatt analysis builds on a Taylor expansion of the estimator and therefore when the bandwidth h_n is large, Rosenblatt's bias analysis tends to provide wrong estimates of the bias, as observed in the experimental section. We consider multidimensional input space, and we apply the bound to a realistic regression problem.

6.1.1. Problem Statement

Consider the problem of estimating $\mathbb{E}[Y|X = \mathbf{x}]$ where $X \sim f_X$ and $Y = m(X) + \epsilon$, with noise ϵ, i.e. $\mathbb{E}[\epsilon] = 0$. The noise can depend on $\mathbf{x}$, but since our analysis is conducted point-wise for a given $\mathbf{x}$, $\epsilon_\mathbf{x}$ will be simply denoted by ϵ. Let $m : \mathbb{R}^d \to \mathbb{R}$ be the *regression function* and f_X a probability distribution on X called *design*. In our analysis we consider $X \in \mathbb{R}^d$ and $Y \in \mathbb{R}$. The Nadaraya–Watson kernel estimate of $\mathbb{E}[Y|X = \mathbf{x}]$ is

$$\hat{m}(\mathbf{x}) = \frac{\sum_{i=1}^n K_\mathbf{h}(\mathbf{x} - \mathbf{x}_i) y_i}{\sum_{j=1}^n K_\mathbf{h}(\mathbf{x} - \mathbf{x}_j)}, \tag{6.1}$$

where $K_\mathbf{h}$ is a kernel function with bandwidth-vector $\mathbf{h}$, the $\mathbf{x}_i$ are drawn from the design f_X and y_i from $m(\mathbf{x}_i) + \epsilon$. Note that both the numerator and the denominator are proportional to Parzen-Rosenblatt density kernel estimates (Rosenblatt, 1956; Parzen, 1962). We are interested in the point-wise bias of such estimate $\mathbb{E}[\hat{m}(\mathbf{x})] - m(\mathbf{x})$. In the prior analysis of Rosenblatt (1969), knowledge of m', m'', f_X, f_X' is required and f, m' must be continuous in a neighborhood of x. In addition, and as discussed in the introduction, the analysis is limited to a one-dimensional design and an infinitesimal bandwidth. We briefly present the classical bias analysis of Rosenblatt (1969) before introducing our results for clarity of exposition.

Distribution	Density	Υ	$\mathcal{D}$	L_f
Laplace(μ, λ)	$\frac{1}{2\lambda} \exp\left(-\frac{|x-\mu|}{\lambda}\right)$	$(-\infty, +\infty)$	$(-\infty, +\infty)$	λ^{-1}
Cauchy$(\mu; \gamma)$	$\left(\pi\gamma + \pi\frac{(x-\mu)^2}{\gamma}\right)^{-1}$	$(-\infty, +\infty)$	$(-\infty, +\infty)$	$\frac{2(z-\mu)}{\gamma^2+(z-\mu)^2}1$
Uniform(a, b)	$\begin{cases} \frac{1}{b-a} & \text{if } a \leq x \leq b \\ 0 & \text{otherwise} \end{cases}$	(a, b)	(a, b)	0
Pareto(α)	$\begin{cases} \frac{\alpha}{x^{\alpha+1}} & \text{if } x \geq 1 \\ 0 & \text{otherwise} \end{cases}$	$(1, +\infty)$	$(1, +\infty)$	$1 + \alpha$
Normal(μ, σ)	$\frac{1}{\sqrt{2\pi\sigma^2}} \exp -\frac{(x-\mu)^2}{2\sigma^2}$	$(-\infty, +\infty)$	(a, b)	$f_{\mu,\sigma}(a, b)$

Table 6.1.: Examples of parameters to use for different univariate random design.

Theorem 4. Classic Bias Estimation (Rosenblatt, 1969). *Let $m : \mathbb{R} \to \mathbb{R}$ be twice differentiable. Assume a set $\{x_i, y_i\}_{i=1}^n$, where $\mathbf{x}_i$ are i.i.d. samples from a distribution with non-zero differentiable density f_X. Assume $y_i = m(\mathbf{s}_i) + \epsilon_i$, with noise $\epsilon_i \sim \varepsilon(\mathbf{x}_i)$. The bias of the Nadaraya–Watson kernel in the limit of infinite samples and for $h \to 0$ and $nh_n \to \infty$ is*

$$\mathbb{E}\left[\lim_{n \to \infty} \hat{m}_n(x)\right] - m(x) = h_n^2 \left(\frac{1}{2}m''(x) + \frac{m'(x)f_X'(x)}{f_X(x)}\right) \int u^2 K(u)\, \mathrm{d}u + o_P\left(h_n^2\right)$$

$$\approx h_n^2 \left(\frac{1}{2}m''(x) + \frac{m'(x)f_X'(x)}{f_X(x)}\right) \int u^2 K(u)\, \mathrm{d}u. \tag{6.2}$$

Note that Equation 6.2 must be normalized with $\int_{-\infty}^{\infty} k(u)\, \mathrm{d}u$ when the kernel function does not integrate to one. The o_P term denotes the asymptotic behavior w.r.t. the bandwidth. Therefore, for a larger value of the bandwidth, the bias estimation becomes worse, as is illustrated in Figure 6.1.

6.2. Main Result

In this section, we present two bounds on the bias of the Nadaraya–Watson estimator. The first one considers a bounded regression function m (i.e., $|m(x)| \leq M$), and allows for weak Lipschitz conditions on a subset of the design's support. Instead, the second bound does not require the regression function to be bounded but only the weak Lipschitz continuity to hold on all of its support. The definition of "weak" Lipschitz continuity will be given below.

To develop our bound on the bias for multidimensional inputs is essential to define some subset of the $\mathbb{R}^d$ space. In more detail, we consider an open n-dimensional interval in $\mathbb{R}^d$ which is defined as $\Omega(\tau^-, \tau^+) \equiv (\tau_1^-, \tau_1^+) \times \cdots \times (\tau_d^-, \tau_d^+)$ where $\tau^-, \tau^+ \in \overline{\mathbb{R}}^d$. We now formalize what is meant by weak (log-)Lipschitz continuity. This will prove useful as we need knowledge of the weak-Lipschitz constants of m and $\log f_X$ in our analysis.

Definition 3. Weak Lipschitz continuity at $\mathbf{x}$ on the set $\mathcal{C}$ under the L_1-norm.
Let $\mathcal{C} \subseteq \mathbb{R}^d$ and $f : \mathcal{C} \to \mathbb{R}$. We call f weak Lipschitz continuous at $\mathbf{x} \in \mathcal{C}$ if and only if

$$|f(\mathbf{x}) - f(\mathbf{y})| \leq L|\mathbf{x} - \mathbf{y}| \quad \forall \mathbf{y} \in \mathcal{C},$$

where $|\cdot|$ denotes the L_1-norm.

Definition 4. Weak log-Lipschitz continuity at $\mathbf{x}$ on the set $\mathcal{C}$ under the L_1-norm.
Let $\mathcal{C} \subseteq \mathbb{R}^d$. We call f weak log-Lipschitz continuous at $\mathbf{x}$ on the set $\mathcal{C}$ if and only if

$$|\log f(\mathbf{x}) - \log f(\mathbf{y})| \leq L|\mathbf{x} - \mathbf{y}| \quad \forall \mathbf{y} \in \mathcal{C}.$$

Note that the set $\mathcal{C}$ can be a subset of the function's domain.

It is important to note that, in contrast to the global Lipschitz continuity, which requires $|f(\mathbf{y}) - f(\mathbf{z})| \leq L|\mathbf{y} - \mathbf{z}|$ $\forall \mathbf{y}, \mathbf{z} \in \mathcal{C}$, the weak Lipschitz continuity is defined at a specific point $\mathbf{x}$ and therefore allows the function to be discontinuous elsewhere. The Lipschitz assumptions are not very restrictive, and in practice require a bounded gradient. They have been widely used in various fields. Note that when the Lipschitz constants are not known, they can be estimated from the dataset Wood & Zhang (1996). In the following we list the set of assumptions that we use in our theorems.

A1. f_X and m are defined on $\Upsilon \equiv \Omega(\mathbf{x} - v^-, \mathbf{x} + v^+)$ and $v^-, v^+ \in \overline{\mathbb{R}}_+^d$;

A2. f_X is log weak Lipschitz with constant L_f at $\mathbf{x}$ on the set $\mathcal{D} \equiv \Omega(\mathbf{x} - \boldsymbol{\delta}^-, \mathbf{x} + \boldsymbol{\delta}^-) \subseteq \Upsilon$ and $f_X(\mathbf{x}) \geq f_X(\mathbf{z})$ $\forall \mathbf{z} \in \Upsilon \backslash \mathcal{D}$ with positive defined $\boldsymbol{\delta}^-, \boldsymbol{\delta}^+ \in \overline{\mathbb{R}}_+^d$ (note that this implies $f_X(\mathbf{y}) > 0$ $\forall \mathbf{y} \in \mathcal{D}$);

A3. m is weak Lipschitz with constant L_m at $\mathbf{x}$ on a the set $\mathcal{G} \equiv \Omega(\mathbf{x} - \boldsymbol{\gamma}^-, \mathbf{x} + \boldsymbol{\gamma}^+) \subseteq \mathcal{D}$ with positive defined $\boldsymbol{\gamma}^-, \boldsymbol{\gamma}^+ \in \overline{\mathbb{R}}_+^d$.

To work out a bound on the bias valid for a wide class of kernels, we must enumerate some assumption and quantify some integrals with respect to the kernel.

A4. The multidimensional kernel $K_\mathbf{h} : \mathbb{R}^d \to \mathbb{R}$ can be decomposed in a product of independent uni-dimensional kernels, i.e., $K_\mathbf{h}(\mathbf{x}) = \prod_{i=1}^d k_i(x/h_i)$ with $k_i : \mathbb{R} \to \mathbb{R}$;

A5. the kernels are non-negative $k_i(x) \geq 0$ and symmetric $k_i(x) = k_i(-x)$;

A6. for every $a, x \in \mathbb{R}$ and $h \neq 0$, the integrals $\int_0^a k_i(x)\,\mathrm{d}x = \Phi_i(a)$, $\int_0^a k_i(x)e^{-xL_f}\,\mathrm{d}x = B_i(x, L_f)$, $\int_0^a k_i(x)xe^{-xL_f}\,\mathrm{d}x = C_i(x, L_f)$ are finite (i.e., $< +\infty$).

Assumptions A4-A6 are not really restrictive, and includes any kernel with both finite domain and co-domain, or not heavy-tailed (e.g., Gaussian-like). Furthermore, Axiom A4 allows any independent composition of different kernel functions. In the Appendix C.3 we detail the integrals of Axiom A6 for different kernels. Note that when the integrals listed in A6 exist in closed form, the computation of the bound is straightforward, and requires negligible computational effort.

In the following, we propose two different bounds of the bias. The first version considers a bounded regression function ($M < +\infty$), this allows both the regression function and the design to be weak Lipschitz on a subset of their domain. In the second version instead, we consider the case of an unbounded regression function ($M = +\infty$) or an unknown bound M. In this case both the regression function and the design must be weak Lipschitz on the entire domain Υ.

Theorem 5. Bound on the Bias with Bounded Regression Function.
Assuming A1–A3, $\mathbf{h} \in \mathbb{R}_+^d$ a positive defined vector of bandwidths $\mathbf{h} = [h_1, h_2, \ldots, h_n]^\mathsf{T}$, K_h the multivariate kernel defined in A4–A6, $\hat{m}_n(\mathbf{x})$ the Nadaraya–Watson kernel estimate using n observations $\{\mathbf{x}_i, y_i\}_{i=1}^n$ with $x_i \sim f_X$, $y_i = m(\mathbf{x}_i) + \epsilon_i$ and with noise $\epsilon_i \sim \varepsilon(\mathbf{x}_i)$ centered in zero ($\mathbb{E}[\varepsilon(\mathbf{x}_i)] = 0$), $n \to \infty$, and furthermore assuming

there is a constant $0 \leq M < +\infty$ such that $|m(\mathbf{y}) - m(\mathbf{z})| \leq M \ \forall \mathbf{y}, \mathbf{z} \in \Upsilon$, the considered Nadaraya–Watson kernel regression bias is bounded by

$$\left| \mathbb{E}\left[\lim_{n\to\infty} \hat{m}_n(\mathbf{x}) \right] - m(\mathbf{x}) \right| \leq \frac{L_m \sum_{k=1}^{d} \xi_k(\phi_k^-, \phi_k^+) \prod_{i\neq k}^{d} \zeta(\phi_i^-, \phi_i^+) + M \left(\prod_{i=1}^{d} \zeta(\gamma_i^-, \gamma_i^+) - \prod_{i=1}^{d} \zeta(\phi_i^-, \phi_i^+) + D \right)}{\prod_{i=1}^{d} \psi_i(\delta_i^-, \delta_i^+)}$$

where

$$\psi_i(a, b) = h_i \left(B_i(b/h_i, L_f h_i) - B_i(-a/h_i, -L_f h_i) \right), \zeta_i(a, b) = h_i \left(B_i(b/h_i, -L_f h_i) - B_i(-a/h_i, L_f h_i) \right),$$

$$\xi_i(a, b) = h_i^2 \left(C_i(b/h_i, -L_f h_i) + C_i(-a h_i, L_f h_i) \right), D = \lim_{\omega \to +\infty} \prod_{i=1}^{d} h_i \Phi_i\left(\frac{\omega}{h_i} \right) - \prod_{i=1}^{d} h_i \varphi_i,$$

with $\varphi_i = \Phi_i(\gamma_i^+/h_i) + \Phi_i(\gamma_i^-/h_i)$, and $0 < \phi_i^- \leq \gamma_i^-, 0 < \phi_i^+ \leq \gamma_i^+$ can be freely chosen to obtain a tighter bound. We suggest $\phi_i^+ = \min(\gamma_i^+, M/L_m), \phi_i^- = \min(\gamma_i^-, M/L_m)$.

In the case where M is unknown or infinite, we propose the following bound.

Theorem 6. Bound on the Bias with Unbounded Regression Function.
Assuming A1–A3, $\mathbf{h} \in \overline{\mathbb{R}}_+^d$ a positive defined vector of bandwidths $\mathbf{h} = [h_1, h_2, \ldots, h_n]^\mathsf{T}$, K_h the multivariate kernel defined in A4–A6, $\hat{m}_n(\mathbf{x})$ the Nadaraya–Watson kernel estimate using n observations $\{\mathbf{x}_i, y_i\}_{i=1}^{n}$ with $x_i \sim f_X$, $y_i = m(\mathbf{x}_i) + \epsilon_i$ and with noise $\epsilon_i \sim \varepsilon(\mathbf{x}_i)$ centered in zero ($\mathbb{E}[\varepsilon(\mathbf{x}_i)] = 0$), $n \to \infty$, and furthermore assuming that $\Upsilon \equiv \mathcal{D} \equiv \mathcal{G}$, the considered Nadaraya–Watson kernel regression bias is bounded by

$$\left| \mathbb{E}\left[\lim_{n\to\infty} \hat{m}_n(\mathbf{x}) \right] - m(\mathbf{x}) \right| \leq \frac{L_m \sum_{k=1}^{d} \xi_k(v_k^-, v_k^+) \prod_{i\neq k}^{d} \zeta_i(v_i^-, v_i^+)}{\prod_{i=1}^{d} \psi_i(v_i^-, v_i^+)},$$

where ξ_k, ζ_i, ψ_i are defined as in Theorem 5.

We detail the proof of both theorems in Appendix C.2. Note that the conditions required by our theorems are mild, and they allow a wide range of random designs, including and not limited to Gaussian, Cauchy, Pareto, Uniform, and Laplace distributions. In general, every continuously differentiable density distribution is also weak log-Lipschitz in some closed subset of its domain. For example, the Gaussian distribution does not have a finite Lipschitz constant on its entire domain, but there is a finite weak Lipschitz constant on any closed interval. Examples of distributions that are weak log-Lipschitz are presented in Table 6.1.

6.2.1. Theoretical Analysis

Although the bound applies to different kernel functions, in the following we analyze the most common Gaussian kernel. It worth noting that for $k(x) = e^{-x^2}$,

$$\phi(x) = \frac{\sqrt{\pi}}{2} \operatorname{erf}(a), \quad B(a, c) = \frac{\sqrt{\pi}}{2} e^{\frac{c^2}{4}} \left(\operatorname{erf}\left(a + \frac{c}{2} \right) - \operatorname{erf}\left(\frac{c}{2} \right) \right),$$

$$C(a, c) = \frac{1}{2} \left(1 - e^{-a(a+c)} - cB(a, c) \right).$$

Note that we removed the subscripts from the functions ψ, B, and C, as we consider only the Gaussian kernel. To provide a tight bound, we consider many quantities that describe the design's domain, the Lipschitz constants of the design and of the regression function, the bound of the image of the regression function, and the different bandwidths for each dimension of the space. This complexity results in an effective but poorly readable bound. In this section, we try to simplify the problem and to analyze the behavior of the bound in the limit.

Asymptotic Analysis: Let us consider, for the moment, the case of one-dimension ($d = 1$) and infinite domains and co-domains (M unknown and $v^- = v^+ = \delta- = \delta+ = \gamma^- = \gamma^+ = \infty$). In this particular case, the bound becomes

$$\left| \mathbb{E}\left[\lim_{n\to\infty} \hat{m}_n(\mathbf{x}) \right] - m(\mathbf{x}) \right| \leq L_m h \left(\frac{1}{2\sqrt{\pi}\exp\frac{L_f^2 h^2}{4}} + hL_f \left(\mathrm{erf}\left(\frac{hL_f}{2}\right) + 1 \right) \right) = A_1.$$

As expected, for $h \to 0$ or for $L_m = 0$, $B_1 = 0$. This result is in line with (6.2) (since $L_m = 0 \implies m' = 0, m'' = 0$). A completely flat design, e.g. uniformly distributed, does not imply a zero bias. This can be seen either in Rosenblatt's analysis or by just considering the fact that, notoriously, the Nadaraya-Watson estimator suffers from the boundary bias. When we analyze our bound, we find in fact that $\lim_{L_f\to 0} B_1 \propto L_m h$. It is also interesting to analyze the asymptotic behavior when these quantities tend to infinity. Similarly to (6.2), we observe that A_1 grows quadratically w.r.t. the kernel's bandwidth h and it scales linearly w.r.t. the Lipschitz constant of the regression function (which is linked to m'). A further analysis brings us to the consideration that Rosenbatt's analysis is linear w.r.t. $\mathrm{d}/\mathrm{d}x \log f_X(x)$ (since $f'_X(x)/f_X(x) = \mathrm{d}/\mathrm{d}x \log f_X(x)$). Our bound has a similar implication, as the Log-Lipschitz constant is also related to the derivative of the logarithm of the design function, and $A_1 = \mathcal{O}(L_f)$.

Boundary Bias: The Nadaraya-Watson kernel estimator is affected by the boundary bias. The boundary bias is an additive bias term affecting the estimation in the region close to the boundaries of the design's domains. Since in our framework, we can consider a closed domain of the design, we can also see what is happening close to the border. Let us consider still a one-dimensional regression, but this time $v^- \to 0$, $v^- = \delta^- = \gamma^-$ and $v^+ = \delta^+ = \gamma^+ = \infty$. In this case, we obtain

$$\left| \mathbb{E}\left[\lim_{n\to\infty} \hat{m}_n(\mathbf{x}) \right] - m(\mathbf{x}) \right| \leq \frac{L_m h}{\sqrt{\pi}e^{\frac{L_f^2 h^2}{4}}\left(1 - \mathrm{erf}\left(\frac{hL_f}{2}\right)\right)} + \frac{L_m L_f h^2\left(1 + \mathrm{erf}\left(\frac{hL_f}{2}\right)\right)}{2 - 2\,\mathrm{erf}\left(\frac{hL_f}{2}\right)} = A_2.$$

Keeping in account that $\mathrm{d}/\mathrm{d}x\,\mathrm{erf}(x) \propto e^{-x}$ and using L'Hôpital's rule, we can observe that the bound is now exponential w.r.t. h and L_f, i.e., $A_2 = \mathcal{O}(e^{hL_f})$, which implies that it is more "sensible" to higher bandwidths or less smooth design. Interestingly, instead, the bounds maintains its linear relation w.r.t. L_m.

Dimensionality: Let us now study the multidimensional case, supposing that each dimension has same bandwidth and same values for the boundaries. In this case,

$$\left| \mathbb{E}\left[\lim_{n\to\infty} \hat{m}_n(\mathbf{x}) \right] - m(\mathbf{x}) \right| \leq \frac{d\xi(v^-, v^+)\zeta(v^-, v^+)^{d-1}}{\psi(v^-, v^+)^d} \propto d\left(\frac{\zeta(v^-, v^+)}{\psi(v^-, v^+)}\right)^{d-1}.$$

Therefore, the bound scales exponentially w.r.t. the dimension. We observe an exponential behavior when $\mathbf{x}$ is close to the boundary of the design's domain. In these regions, in fact the ratio $\zeta(v^-, v^+)/\psi(v^-, v^+)$ is particularly high. Of course, when the aforementioned ratio tends to one, the linearity w.r.t. d is predominant.

We can conclude the analysis by noticing that our bound has similar limiting behavior with the Rosenblatt's analysis, but it provides a hard bound on the bias.

6.3. Empirical Analysis

In this section, we provide three numerical analyses of our bounds on the bias[2]. The first analysis of our method is conducted on uni-dimensional input spaces for display purposes and aims to show the properties of our bounds in different scenarios. The second analysis aims instead at testing the behavior of our method on a multidimensional input space. The third analysis emulates a realistic scenario where our bound can be applied.

Uni-dimensional Analysis: We select a set of regression functions with different Lipschitz constants and different bounds,

- $y = \sin(5x)$; $L_m = 5$ and $M = 1$,

- $y = \log x$ which for $\mathcal{G} \equiv \Omega(-1, +\infty)$ has $L_m = 1$ and $M = +\infty$,

- $y = 60^{-1}\log\cosh 60x$ which has $L_m = 1$, is unbuounded, and has a particularly high second derivative in $x = 0$, with $m''(0) = 60$,

- $y = \sqrt{x^2 + 1}$ which has $L_m = 1$ and is unbounded.

A zero-mean Gaussian noise with standard deviation $\sigma = 0.05$ has been added to the output y. Our theory applies to a wide family of kernels. In this analysis we consider a Gaussian kernel, with $k(x) = e^{-x^2}$, a box kernel, with $k(x) = I(x)$, and a triangle kernel, with $k(x) = I(x)(1 - |x|)$ with

$$I(x) = \begin{cases} 1 & \text{if} |x| \leq 1, \\ 0 & \text{otherwise.} \end{cases}$$

We further analyze the aforementioned kernels in Appendix C.3. In order to provide as many different scenarios as possible we also used the distributions from Table 6.1, using therefore both infinite domain distributions, such as *Cauchy* and *Laplace*, and finite domain such as *Uniform*. In order to numerically estimate the bias, we approximate $E[\hat{m}_n(x)]$ with an ensemble of estimates $N^{-1}\sum_{j=1}^{N} \hat{m}_{n,j}(x)$ where each estimate $\hat{m}_{n,j}$ is built on a different dataset (drawn from the same distribution f_X). In order to "simulate" $n \to \infty$ we used $n = 10^5$ samples, and to obtain high confidence of the bias' estimate, we used $N = 100$ models.

In this section we provide some simulations of our bound presented in Theorem 5 and Theorem 6, and for the Rosenblatt's case we use

$$\left| h_n^2 \left(\frac{1}{2}m''(x) + \frac{m'(x)f'_X(x)}{f_X(x)} \right) \int u^2 K(u)\,\mathrm{d}u \right|.$$

Since the Rosenblatt's bias estimate is not an upper bound, it can happen that the true bias is higher (as well as lower) than this estimate, as it is possible to see in Figure 6.1. We presented different scenarios, both with bounded and unbounded functions, infinite and finite design domains, and a larger or smaller bandwidth choice. It is possible to observe that, thanks to the knowledge of f, f', m', m'' the Rosenblatt's estimation of the bias tends to be more accurate than our bound. However, it can happen that it largely overestimates the bias, like in the case of $m(x) = 60^{-1}\log\cosh(60x)$ in $x = 0$ or to underestimate it, most often in boundary regions. In contrast, our bound always overestimates the true bias, and despite its lack of knowledge of f, f', m', m'', it is most often tight. Moreover, when the bandwidth is small, both our method and Rosenblatt's deliver an accurate estimation of the bias. In general, Rosenblatt tends to deliver a better estimate of the bias, but it does not behave as a bound, and in some situations, it also can deliver larger mispredictions. In detail, the plot (a) in Figure 6.1 shows that with a tight bandwidth, both our method and Rosenblatt's method achieve good approximations of the

---

[2]The code of our numerical simulations can be found at http://github.com/SamuelePolimi/UpperboundNWBias.

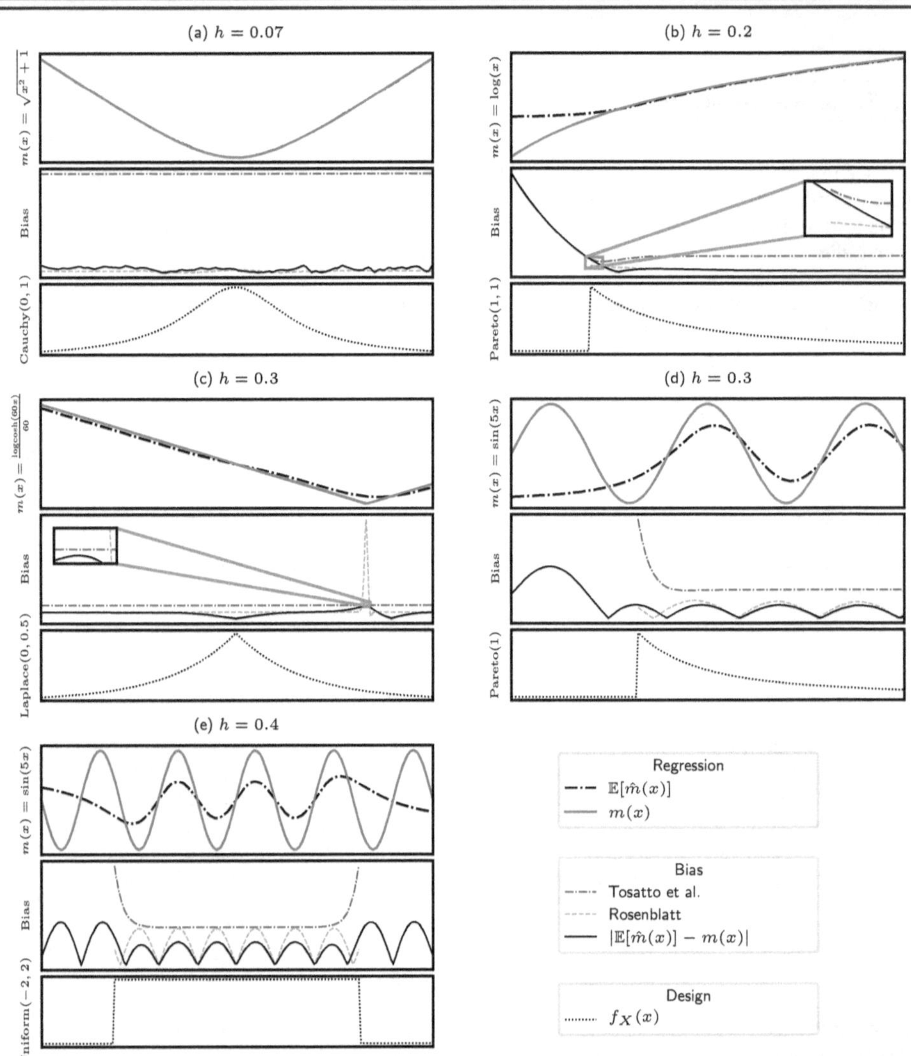

Figure 6.1.: We propose some simulations of Nadaraya–Watson regression with different designs, regression functions and bandwidths. The regression function $m(x)$ is represented with a solid line, while the Nadaraya–Watson estimate $\hat{m}(x)$ is represented with a dash-dotted line in the top subplot of each experiment. In the second subplots, it is possible to observe the true bias (solid line), as well as our upper bound (dashed line) and Rosenblatt's estimate (dash-dotted line). In the bottom subplots depict the design used. The bandwidth used for the estimation is denoted with h. It is possible to observe that Rosenblatt's estimate often under or overestimates the bias. In all the different test conditions, our method correctly upper bounds the bias.

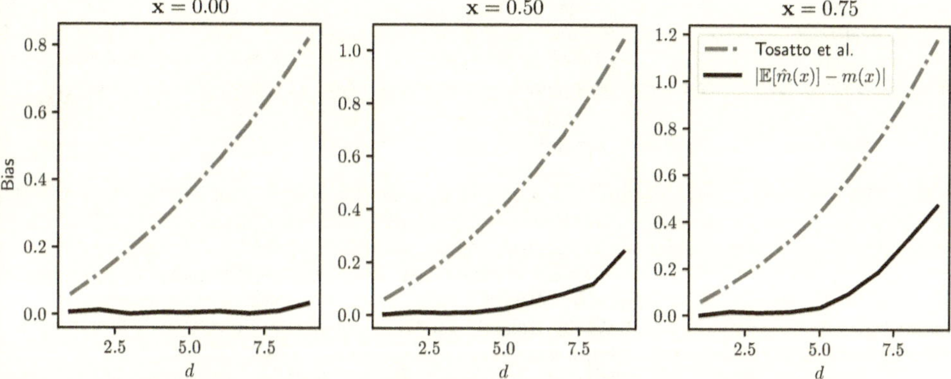

Figure 6.2.: We propose a study on how the bias vary w.r.t. the dimension. We can see that our bound is approximately linear far from the boundary, while close to the bound, it becomes exponential. The same behavior is also observed with the "true bias" (computed via numerical simulations).

bias, but only our method correctly upper bounds the bias. When increasing the bandwidth, we obtain both a larger bias and subsequent larger estimates of the bias. Our method consistently upper bounds the bias, while in many cases, Rosenblatt's method underestimates it, especially in proximity of boundaries (subplots b, d, e). An interesting case can be observed in subplot (c), where we test the function $m(x) = 60^{-1} \log \cosh(60x)$, which has a high second-order derivative in $x = 0$: in this case, Rosenblatt's method largely overestimates the bias.

The figure shows that our bound can deal with different functions and random designs, being reasonably tight, if compared to Rosenblatt's estimation, which requires the knowledge of the regression function and the design, and respective derivatives. **Multidimensional Analysis:** We want to study if our bounds work in a multidimensional case and how much it overestimates the true bias (therefore, how tight it is). For this purposes, we took a linear function $m(\mathbf{x}) = \mathbf{1}^\top \mathbf{x}$ where $\mathbf{1}$ is a column-vector of d ones. This function, for any dimension d, has a Lipschitz constant $L_m = 1$ and is unbounded ($M = \infty$). We set a Gaussian design with zero mean and unit diagonal covariance. Since in higher dimensions, the estimation's variance grows exponentially (Györfi et al., 2006), we used a large number of samples

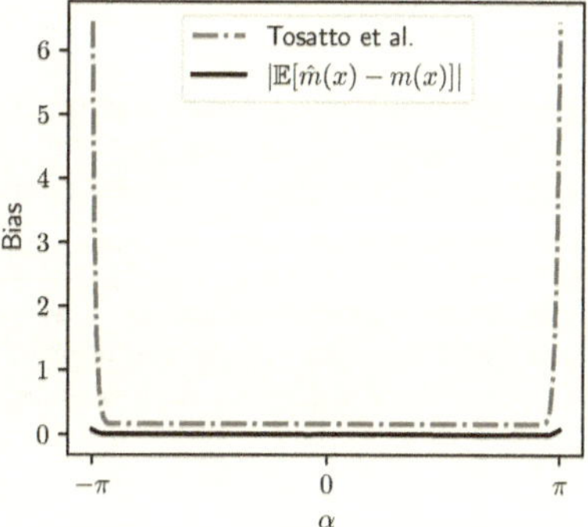

Figure 6.3.: *Experiment on the underactuated pendulum. Our bound gives the possibility to ensure that $\mathbb{E}[m(x) - \hat{m}(x)]$ does not exceed a certain quantity. In this example, it is possible to observe that the bias is correctly bounded.*

($n = 10^6$), and we averaged over $N = 10^5$ independent estimations. In Figure 6.2 we show how the "true" bias (estimated numerically averaging over a thousand Nadaraya-Watson regressions) and our bound evolve with a growing number of dimensions d. Far from the low-density region $\mathbf{x} = 0$ we notice that the bias tends to have a linear behavior, while close to the boundary the bias tends to be exponential. We can observe that our bound correctly bounds the bias in all the cases.

Realistic Scenario: Let us consider the regression problem of the dynamics of an under-actuated pendulum of length l and mass m. In particular, the state of the pendulum can be described by its angle α and its angular rotation $\dot{\alpha}$. Furthermore, a force u can be applied to the pendulum. The full system is described by,

$$\ddot{\alpha} = \frac{3}{ml^2}u - \frac{3g}{2l}\sin(\alpha + \pi),$$

where g is the gravitational acceleration. In practice, when this model is discretized in time, the next state is estimated via numerical integration. Fort this reason, the Lipschitz constant L_m is unknown. Notice that also m' and m'' required by the Rosenblatt's analysis are unknown. We estimated the Lipschitz constant L_m by selecting the highest ratio $|y_i - y_j|/|\mathbf{x}_i - \mathbf{x}_i|$ in the dataset. In our analysis, we want to predict all the states with fixed $\dot{\alpha} = 0$, $u = 0$, but variable $\alpha \in [-\pi, \pi]$. In order to generate the dataset, we use the simulator provided by **gym** (Brockman et al., 2016). To train our models, we generate tuples of $\alpha, \dot{\alpha}, u$ by sampling independently each variable from a uniform distribution i.e., $\alpha \sim \text{Uniform}(-\pi, \pi)$, $\dot{\alpha} \sim \text{Uniform}(-8, 8)$ and $u \sim \text{Uniform}(-2, 2)$ (hence, $L_f = 0$). We fit 100 different models with 50000 samples. We choose a Gaussian kernel with bandwidth $\mathbf{h} = [0.2, 0.2, 0.2]$. Figure 6.3 depicts our bound and the estimated bias. We notice the bias is low and increases close to the boundaries. Our upper bound is tight, but, as expected, it becomes overly pessimistic in the boundary region.

6.4. Epilogue

The Nadaraya–Watson kernel regression is one of the most well-known nonparametric estimator, used in a wide range of applications. Its asymptotic bias and variance are well-known in the literature. However, to the best of our knowledge, such an estimator's bias has never been bounded before. In this paper, we proposed a hard bound of the bias that requires mild assumptions. Our proposed bound is numerically tight and accurate for a large class of regression functions, kernels, and random designs. We believe that providing hard, non-probabilistic guarantees on a regression error is an essential step in adopting data-driven algorithms in real-world applications. Our future research will focus on extending the bias analysis to a broader class of kernel functions and with a finite-samples analysis.

7. Exploration Driven By an Optimistic Bellman Equation

In the previous chapters, we examined dimensionality reduction for robotic movements, off-policy gradient estimates, and we equipped the theory with guarantees on the estimator involved. We implicitly assumed that our data was good enough to allow the reinforcement learning system to find a satisfying solution. However, in a complex scenario, the reinforcement learning algorithm may fail. One of the reasons is that the algorithm can fall in bad local optima. To avoid this, one can design a highly informative reward function that helps the algorithm skip local optima. This approach requires knowledge of the task. Equipping the reinforcement learning algorithm with good exploration can help in skipping those local optima without human intervention. In this chapter, we provide an optimistic Bellman equation derived from information-theoretic principles. Our equation propagates optimistic bonuses in the Bellman recursion allowing a far-sighted exploration, resulting particularly effective for simple, sparse reward signals.

7.1. Prologue

In recent years, the research on reinforcement learning (RL) has made enormous advances, especially on high-dimensional tasks, such as Atari games (Mnih et al., 2015). One of the open problems in such complex domains is how to explore the environment in order to uncover sparse valuable states. Before seeing these states the agent does not necessarily have much information to base its decisions on. For example, an agent may always perceive a null reward except for a terminal state that is particularly difficult to reach. Before the agent reaches the terminal state and observes the subsequent reward, it cannot connect its actions to rewards. In this particular case, the traditional quest for solving the exploration/exploitation trade-off (near)-optimally makes no sense since the agent has no information to reason about possible rewards that it has not yet observed.

A classical way to solve the exploration problem is to explore randomly. Classical approaches such as ϵ-greedy may fail as the probability of reaching the positive reward can be low. A more effective strategy should take into account the underlying uncertainty and try to minimize it, in order to maximize the information gain. Bayesian approaches consider the uncertainty in a principled way but are often computational demanding (Vlassis et al., 2012; Engel et al., 2005). Recently, computationally feasible algorithms inspired by Bayesian principles have been introduced (Azizzadenesheli et al., 2018; Osband et al., 2016). Bootstrapped DQN (BDQN) (Osband et al., 2016) uses an ensemble of value functions in order to have different estimates of the Q-value function approximating posterior sampling (Strens, 2000).

To the best of our knowledge, there is no algorithm among these approximate techniques that is particularly suited for very sparse rewards in high dimensional state space. Our hypothesis is that Bayesian methods are in general more focused on balancing between exploration and exploitation while they cannot achieve deep exploration. The broad category of algorithms based on *intrinsic motivation* (IM) (Chentanez et al., 2004), have less theoretical guarantees than Bayesian approaches, yet they have obtained impressive results for example in the challenging Montezuma's Revenge task (Bellemare et al., 2016). IM algorithms define an additional *intrinsic* reward, which acts as an exploration bonus. Often, the additional reward is defined using heuristics, such as counting state visits and rewarding less visited states (Ostrovski et al., 2017), or by "surprise", that is, the error in predicting future states (Pathak et al., 2017). The drawback of IM techniques is their lack of a principled definition of the intrinsic reward for exploration.

Another related class of techniques is based on optimism; which provides an optimistic estimation under uncertainty, encouraging in this way exploration of uncertain region. Optimism can be categorized in 1) *optimistic initial values*, where the main concept is to initialize the Q-value function with high values in order to ensure enough exploration (Sutton & Barto, 2018; Even-Dar & Mansour, 2002); and 2) methods that require *confidence interval estimation*, such as IEQL+ (Meuleau & Bourgine, 1999; Kumaraswamy et al., 2018) which directly estimates Q-value confidence intervals.

Contribution We introduce a novel Optimistic Bellman Equation (OBE). The OBE results in an optimistic Q-value estimate from an ensemble of value functions where the optimistic estimate is obtained from a maximum-entropy principle. For the exploration bonus that OBE implicitly defines, we can prove that the bonus decreases consistently with the number of state visits. Our proposed algorithm can be seen as a mixture of different techniques: as an approximated Bayesian method, we estimate the uncertainty with an ensemble; like optimism-based methods, we select optimistic estimates, and like IM, we propagate an implicit exploration bonus. Nevertheless, OBE can be applied to a wide range of algorithms by introducing an ensemble for the estimation of the critic (as done in (Osband et al., 2016), (Chen et al., 2017)) and a softmax for updating the entries.

7.1.1. Problem Statement

An infinite-horizon discounted time-discrete Markov Decision Process (MDP) is defined by a tuple of $< \mathcal{S}, \mathcal{A}, R, P, \gamma, \mu_0 >$ where $\mathcal{S}$ is the set of the states, $\mathcal{A}$ is a finite set of actions, $R : \mathcal{S} \times \mathcal{A} \to \mathcal{M}(\mathbb{R})$ where $\mathcal{M}(\mathcal{Z})$ denotes the sets of probability measures over the space Z, $P : \mathcal{S} \times \mathcal{A} \to \mathcal{M}(\mathcal{S})$ is the transition distribution, $\gamma \in [0, 1)$ denotes the discount factor and μ_0 is the initial state distribution. We define the set of deterministic policies as $\Pi : \mathcal{S} \to \mathcal{A}$. Our goal is to find the optimal policy $\pi^* \in \Pi$ that maximizes the expected return

$$J^\pi = \mathbb{E}\left[\sum_{t=1}^{T} \gamma^{(t-1)} r_t\right],$$

where $r_t \sim R(s_t, a_t)$, $s_t \sim P(\cdot|s_{t-1}, a_{t-1})$, $a_t = \pi(s_t)$, $s_1 \sim \mu_0$. A common approach to solve such a problem is to find the so-called optimal Q-value function Q^*, which is the solution to the optimal Bellman equation (BE):

$$Q^*(s, a) = \overline{R}(s, a) + \gamma \sum_{s' \in \mathcal{S}} P(s'|s, a) \max_{a' \in \mathcal{A}} Q^*(s', a'), \tag{7.1}$$

where $\overline{R}(s, a) = \mathbb{E}[R(s, a)]$ and the summation could be replaced by an integral, depending on whether $\mathcal{S}$ is discrete or continuous. Once the optimal Q-value function Q^* is found, we know that the optimal policy is equivalent to $\pi^*(s) = \arg\max_a Q(s, a)$.

The algorithm we propose could be considered as a mixture of different techniques: as an approximated Bayesian method (Engel et al., 2005; Vlassis et al., 2012), we estimate the uncertainty with an ensemble Osband et al. (2016); like optimism (Lai et al., 1995; Kearns & Singh, 2002; Brafman & Tennenholtz, 2002; Azizzadenesheli et al., 2018), we select optimistic estimates, and like intrinsic motivation (IM) approaches (Chentanez et al., 2004; Schmidhuber, 2008; White & White, 2010), we propagate an implicit exploration bonus. Unlike many IM techniques, our bonus is defined implicitly in a so-called *optimistic Bellman equation* (OBE), by selecting an optimistic estimate from an ensemble of value functions.

Intrinsic motivation algorithms define an additional *intrinsic* reward often using heuristics, such as counting state visits and rewarding less visited states (Ostrovski et al., 2017; Bellemare et al., 2016), or by "surprise", that is, the error in predicting future states (Pathak et al., 2017). Approaches based on the optimism in the face of uncertainty principle (Lai et al., 1995; Kearns & Singh, 2002; Brafman & Tennenholtz, 2002) add an additional reward term to state-action pairs proportional to the amount of uncertainty. The amount of uncertainty, and thus the additional reward, usually depends on the amount of information collected of a state-action pair. Due to explicit uncertainty modeling these methods are able to restrict exploration to regions where the policy is still far from the optimal solution. Often these methods provide performance guarantees. However, in practice the method may not always converge (Jaksch et al., 2010; Osband et al., 2016). Bayesian posterior sampling has shown performance improvements over optimism in the face of uncertainty methods (Strens, 2000; Osband et al., 2013; Grande et al., 2014).

There is a broad range of recent work on exploration in high dimensional state-spaces. UCB Exploration via Q-Ensembles (Chen et al., 2017), similarly to our approach, uses M different estimates of the Q value function in order to infer the uncertainty of the estimate. In particular, Chen et al. (2017) use an optimistic bound for the policy:

$$a \in \arg\max_{a} \mu(s, a) + \lambda\sigma(s, a).$$

compute directly an estimate of the standard deviation over multiple estimations of the Q function, and use it to guide the exploration. However, such methods do not propagate the variance through the Bellman equation, and thus the agent is not able to be foresighted w.r.t. future exploration possibilities. Interestingly, Kumaraswamy et al. (2018) propose a context dependent upper confidence estimation based on least square temporal difference, providing less computational demand than fully Bayesian approaches, and better performance than ensemble-based methods.

Contrary to these approaches, the recent "uncertainty Bellman equation" (UBE) (O'Donoghue et al., 2018) propagates variance estimates of the Q-value with a Bellman recursion and uses the estimates for posterior sampling (our action selection is deterministic). For tabular policies, UBE estimates local uncertainty proportional to the inverse visitation count similarly to count based approaches and for neural network policies a linear uncertainty approximation is used. Instead, our implicit state/action specific uncertainty estimation is based on the diversity in the Q-function ensemble fwhile taking future Q-function uncertainty into account. Moreover, contrary to our approach, UBE assumes that the MDP is a directed acyclic graph.

Entropic regularization has been extensively used in reinforcement learning, either for ensuring stability (Peters et al., 2010; Schulman et al., 2015; Mnih et al., 2016) and sample efficiency or for providing risk awareness (Howard & Matheson, 1972; Marcus et al., 1997; Ruszczyński, 2010). However, the entropic regularization is usually performed on the state-action distribution (Neu et al., 2017), and thus applied to the so-called "aleatoric" uncertainty. This kind of uncertainty is related to the intrinsic stochasticity of the MDP and to the state-action distribution induced by the policy. The Boltzmann policy (Sutton et al., 2000;

Kaelbling et al., 1996) or the soft Bellman equations are derived from this principle. In contrast, our entropic regularization is applied to the "epistemic" uncertainty, or, in other words, to the model uncertainty (in this specific case to the Q-function uncertainty). To the best of our knowledge this is the first attempt to use entropic regularization on the epistemic uncertainty to drive exploration.

7.2. Learning Value Function Ensembles with Optimistic Estimate Selection

Ensemble methods (Opitz & Maclin, 1999) are a prevalent machine learning technique where multiple models are used to learn the same target function. In addition to being commonly used to improve the generalization of the prediction, ensemble methods offer a simple way to estimate the uncertainty of the prediction. We consider the application of ensemble methods in the RL framework with the purpose of approximating the action-value functions. Indeed we want to obtain a cheap estimate of the uncertainty of action-values, in order to apply the *optimism in the face of uncertainty* (OFU) principle in action selection.

7.2.1. An Optimistic Bellman Equation for Action-Value Function Ensembles

The core of our work consists of a BE for action-value ensembles which incorporates the information about the uncertainty provided by a Q-function ensemble. In more detail, we want to overestimate the action-value functions with the result of encouraging exploration. Thus, we propose an optimistic Bellman equation (OBE) which propagates an optimistic estimate of the action-value function. We want to emphasize that when all the Q-functions of the ensemble are identical, we assume that there is no uncertainty, and under this condition the OBE will behave exactly equivalently to the classic BE. The solution Q^* of OBE is the same of the classic BE. In other words, the OBE differs from the classic BE when it is not satisfied, and more precisely when approximation is introduced either by limited availability of samples and/or functional approximation. This

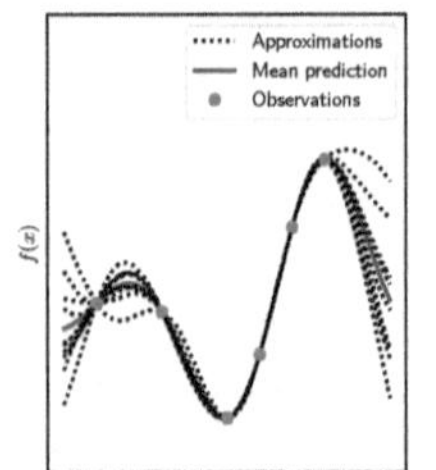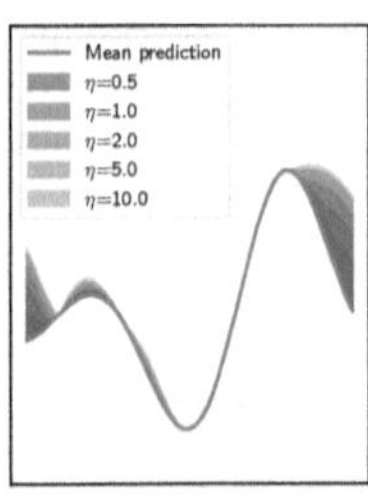

Figure 7.1.: **Left:** Different estimates of a function. **Right:** The entropic-map combines the function estimates to obtain an optimistic estimate where η controls the level of optimism.

makes sense, since when the perfect solution is available there is no need for optimism and exploration. The optimistic Bellman operator derived from the OBE, enjoys the classical properties, like contractivity and the existence of a unique fixed point, enabling its usage in value-based or actor-critic reinforcement algorithms. The diversity in the Q-value ensemble should be ideally consistent with the uncertainty of the estimation; e.g. when the estimate is certain, all the values in the ensemble should agree on the same value, otherwise the ensemble should have discordant values.

Given an ensemble of Q-value functions $\{Q_m\}_{m=1}^{M}$, we want to work out an optimistic estimate from the diverse estimates provided by the ensemble. The simplest and most optimistic solution is to select the highest value

$$\max_{m} Q_m(s, a).$$

However selecting the highest estimate makes poor use of the information provided by the ensemble and can overfit the noise. In order to mitigate this effect, we introduce a notion of *belief* over the estimates where $b_m(s,a)$ is the belief of $Q_m(s,a)$. The main idea is to add an entropic regularization term to the objective (i.e., $\max_{b(s,a)} \sum_m b_m(s,a)Q_m(s,a) - c\sum_m b_m \log b_m$); or to bound the information loss (i.e., $-\sum_m b_m \log b_m \geq \psi$). Hard constraint on the information loss is more appealing since the introduced hyper-parameter does not depend on the magnitude of the rewards but has no closed-form solution. In contrast, the penalization weighting constant introduced by the soft-constraint regularization term is sensitive to the magnitude of the rewards, but admits a closed-form solution. We define two different problems where we use an optimistic estimate of the Q-value function.

Entropy-Regularized Optimistic Q Selection. We define here a Bellman equation over the Q-function ensemble by introducing an optimistic estimate penalized by an entropic regularization term.

Problem 1 (Regularized version).

$$Q_i(s,a) = \max_{b(s,a)\in\mathcal{P}^M} f\big(s,a;b(s,a)\big) - \tfrac{1}{\eta}D_{\mathrm{KL}}\big(b(s,a)\|u\big)$$

$$\text{s.t. } \textstyle\sum_{m=1}^M b_m(s,a) = 1$$

$$\forall s,a,i \in \mathcal{S} \times \mathcal{A} \times \{1,\ldots,M\}$$

where $f(a,s;p) = R(s,a) + \gamma\sum_m b_m(s,a)Q'_m(s,a)$, $Q'_m(s,a) = \sum_{s'} P(s'|s,a)\max_{a'} Q_m(s',a')$, $u_m = 1/M$, $D_{KL}(b(s,a)\|u)$ *is the Kullback-Leibler divergence between the belief $b(s,a)$ and the uniform distribution u.*

Notice that we define all the Q_i to be exactly identical, therefore the solution of Problem 1 admits a solution where all the Q-functions Q_i are identical. However, we will see that the derivation of the optimistic Q-learning (Definition 12) will keep diversity between the ensemble, provided that the Q-function are initialized differently. The algorithm will eventually converge to the unbiased solution of all identical Q-functions as the uncertainty decreases over time.

The choice of using the relative entropy instead of the absolute one has two main advantages: it admits a solution for $\eta \to 0$ and provides a normalization factor. Since problem definition 1 is a convex constrained problem, it is solvable by dual optimization. Introducing λ as Lagrangian multiplier for the constraint, we write the Lagrangian

$$L_i(s,a) \;=\; f(s,a;b(s,a)) - \frac{1}{\eta}D_{\mathrm{KL}}\big(b(s,a)\|u\big) + \lambda\bigg(\sum_m b_m(s,a) - 1\bigg). \tag{7.2}$$

Requiring the partial derivatives of L_i w.r.t b_m and λ to be zero yields to

$$b_m(s,a) = \frac{e^{\eta\gamma Q'_m(s,a)}}{\sum_{k=1}^M e^{\eta\gamma Q'_k(s,a)}}. \tag{7.3}$$

By substituting b_m in (7.2), we obtain the solution to the problem (a detailed derivation is provided in Appendix D (Tosatto et al., 2018))

Definition 5. [1]

$$Q_i(s,a) = \begin{cases} \overline{R}(s,a) + \frac{1}{\eta} \log \frac{\sum_{m=1}^{M} e^{\eta \gamma Q'_m(s,a)}}{M} & \text{if } \eta \neq 0 \\ \overline{R}(s,a) + \frac{\gamma}{M} \sum_{m=1}^{M} Q'_m(s,a) & \text{otherwise} \end{cases} \tag{7.4}$$

Notice that $\eta > 0$ leads to a positive (optimistic) biased estimation, while $\eta < 0$ will leads to a negative (pessimistic) estimate; in this work we will always assume $\eta > 0$ (and therefore we refer to the equation as optimistic). However, in general, the choice of η is difficult since it depends on the magnitude of the reward function. For this reason we introduce the constrained version of the proposed problem.

Optimistic Q Selection Bounding the Information Loss. We bound the information loss between the distribution b_m and the uniform distribution to maintain compatibility with Problem 1. The information loss is bounded between $-\log M$ and 0 where $-\log M$ stands for complete information loss (i.e., only one model is selected) while 0 corresponds to no information loss (i.e., uniform belief distribution). Constraining the information loss has succeeded in prior work, for instance in policy search methods such as (Peters et al., 2010).

Problem 2 (Constrained version).

$$\begin{aligned} Q_i(s,a) = \max_{b(s,a) \in \mathcal{P}^M} \quad & f(s,a;b(s,a)) \\ \text{s.t.} \quad & D_{\mathrm{KL}}\big(b(s,a)\|u\big) \leq \iota_{\max} \\ & \sum_{m=1}^{M} b_m(s,a) = 1 \\ \forall s,a,i \in \mathcal{S} \times \mathcal{A} \times & \{1,\dots,M\} \end{aligned}$$

Using the Lagrangian multiplier β associated with the KL constraint, we obtain the Lagrangian

$$L_i = f(s,a;b(s,a)) + \beta\Big(D_{\mathrm{KL}}\big(b(s,a)\|u\big) - \iota_{\max}\Big) + \lambda\Big(\sum_m b_m(s,a) - 1\Big). \tag{7.5}$$

Substituting β with $-1/\eta$ we note that (7.5) becomes identical to (7.2) except for a constant factor. Since we can not solve η (or β) analytically, we obtain an approximate solution by iteratively optimizing η (or β) and b_m subsequently. OBE takes its name from the fact that when $\eta > 0$, the *log-sum-exp* acts as a *soft-max* operator. Such operator is also well known as an *entropic mapping*, as it can be derived from a maximum-entropy principle. Figure 7.1 shows how the entropic mapping works. The use of the entropic mapping is not new in reinforcement learning: (Asadi & Littman, 2017) propose an interesting use of the entropic mapping as a soft-max over the action in the Bellman equation; (Peters et al., 2010) instead obtain it from an entropic regularization over the state-action distribution. However, as we discussed in Section 7.1.2, differently from the mentioned approaches, our entropic regularization is applied to the epistemic uncertainty.

---

[1] We extend the solution for $\eta = 0$ by computing the limit.

Relation to Intrinsic Motivation. In order to highlight the connection between OBE and IM, we reformulate OBE utilizing the unbiased average of the estimates instead of the log-sum-exp, which includes the positive bias

$$Q_i(s,a) \;=\; \overline{R}(s,a) + U(s,a) + \gamma \sum_{m=1}^{M} \frac{Q'_m(s,a)}{M} \tag{7.6}$$

with the resulting exploratory bonus U

$$U(s,a) = \frac{1}{\eta} \log \sum_{m=1}^{M} \frac{e^{\eta\gamma Q'_m(s,a)}}{M} - \gamma \sum_{m=1}^{M} \frac{Q'_m(s,a)}{M}. \tag{7.7}$$

As $\sum_{i=1}^{N} e^{\eta x_i}/N$ is the *moment generator* w.r.t. samples $\{x_i\}_{i=1}^{N}$, we can rephrase the exploration bonus as

$$U(s,a) = \lim_{N\to+\infty} \frac{1}{\eta} \log\left[1 + \sum_{n=2}^{N} \frac{(\eta\gamma)^n}{n!} \mathcal{M}_n(s,a) \right] = \eta\gamma\mathcal{M}_2(s,a) + O(\eta^2) \tag{7.8}$$

where $\mathcal{M}_n$ is the nth central moment of the random variable Q'_m (Proof in Appendix D)

$$\mathcal{M}_n(s,a) = M^{-1} \sum_{m=1}^{M} \left[\left(Q'_m(s,a) - \overline{Q}'(s,a) \right)^n \right]$$

with

$$\overline{Q}'(s,a) = M^{-1} \sum_{m=1}^{M} Q'_m(s,a).$$

Equation 7.6 shows that OBE is equivalent to BE with an additional bonus defined by Equation 7.8. The bonus U (for any positive η) is always positive, and provides a measure of the uncertainty w.r.t. Q. This is why OBE can be interpreted as a special case of IM.

Explicit Exploration. A general problem affecting intrinsically motivated algorithms, is that the policy greedy to the obtained Q-value function, is not optimized for the original problem. As a solution to this issue we approximate two functions: $\tilde{Q}$, which will be updated using the true reward and Q_E which will be updated using only the intrinsic reward (Szita & Lorincz, 2008). In this way we obtain both the intrinsically motivated policy $\pi_o(s) = \arg\max_a \tilde{Q}(s,a) + Q_E(s,a)$ and the classic policy $\pi_u(s) = \arg\max_a \tilde{Q}(s,a)$. Define

$$\tilde{Q}_i(s,a) = R(s,a) + \gamma \sum_{m=1}^{M} \frac{\tilde{Q}'_m(s,a)}{M} \tag{7.9}$$

with

$$\tilde{Q}'_m(s,a) = \sum_{s'} P(s'|s,a) \max_{a'} \tilde{Q}_m(s',a') \tag{7.10}$$

to obtain an unbiased estimate of the Q-value function, yielding

$$\begin{aligned}
Q_E(s,a) \;&=\; \sum_{t=0}^{T} \gamma^t U(s_t, a_t) | s_0 = s, a_0 = a \\
&=\; \eta^{-1} \log \frac{\sum_{k=1}^{M} e^{\eta\gamma \max_{a'} \tilde{Q}_k(s',a') + Q_E(s',a')}}{M} - \frac{\sum_{k=1}^{M} \gamma \max_{a'} \tilde{Q}_k(s',a')}{M}.
\end{aligned} \tag{7.11}$$

By a simple equation rearrangement, it is possible to show that $\tilde{Q}_i(s,a) + Q_E(s,a)$ is equivalent to $Q_i(s,a)$ as defined in the OBE (7.4).

7.2.2. Optimistic Value Function Estimators

The OBE offers a theoretical framework in which it is possible to develop optimistic value based algorithms. In fact, OBE enjoys all the desirable properties of the BE (e.g. max-norm contractivity), as shown in Appendix D. We present briefly two practical applications of the OBE, an optimistic variant of Q-learning (OQL) and deep Q-network (ODQN).

Optimistic Q-Learning. Motivated by the idea of employing an ensemble of regressors as is done in boostrapped DQN (BDQN) (Osband et al., 2016), we assume to have M randomly initialized Q-tables. Inspired by the well known Q-learning update rule, we derive an optimistic version which is consistent with the OBE.

Definition 6 (Optimistic Q-learning). [2]

$$Q_{i,t+1}(s,a) = (1 - \alpha_t)Q_{i,t}(s,a)l + \alpha_t \left(r_t + \frac{1}{\eta} \log M^{-1} \sum_{j=1}^{M} e^{\gamma \max_{a'} Qj,t(s',a')} \right).$$

We show that, with the update rule proposed, given infinite visits of each state-action pair, all the tables will converge to the same values, and more precisely, after each update, the nth central moment of the updated cell is scaled exactly by $(1 - \alpha_t)^n$:

$$\mathcal{M}_{n,t+1}(s,a) = (1 - \alpha_t)^n \mathcal{M}_{n,t}(s,a) \tag{7.12}$$

where

$$\mathcal{M}_{n,t}(s,a) = M^{-1} \sum_{i=1}^{M} \left(Q_{i,t}(s,a) - \sum_{k=1}^{M} \frac{Q_{k,t}(s,a)}{M} \right)^n.$$

This implies that a cell updated N times, with learning rates $\{\alpha_i\}$, will have the nth central moments scaled by $\Pi_{\alpha_i}(1 - \alpha_i)^n$ w.r.t. the initial one. This leads us to some interesting considerations: 1) the bonus decrease accordingly to the number of state visits; 2) differing from several count-based approaches, our algorithm takes into account the impact of the learning rate; 3) in the limit of an infinite number of visits, the exploration bonus converges to zero. Further details, including a proof of convergence, are given in Appendix D[3]. All the considerations done so far provide a deeper insight about how the algorithm works and its properties. However, in a more complex settings, (e.g., function approximation) the convergence to zero of the exploratory bonus is not guaranteed in general.

Optimistic DQN. In addition to the novel OQL algorithm described previously that can be used for limited discrete state spaces, we propose another algorithm for continuous state spaces based on our OBE. We take inspiration from the framework provided by BDQN (Osband et al., 2016) that uses an ensemble of neural networks as estimator for the Q value function. BDQN minimizes the loss

$$\mathcal{L}_B(s,a) = \sum_{k=1}^{M} \left(r + \gamma \max_{a'} Q_k^T(s',a') - Q_k(s,a) \right)^2,$$

---

[2] We use α_t as a shortcut for $\alpha_t(s,a)$.

[3] We based our convergence proof for OQL on (Melo, 2001) and (Jaakkola et al., 1994)

Algorithm 3 Optimistic DQN

Input: $\{Q_k\}_{k=1}^K$, $\iota_{\max}$, η_{init}, χ, N, C
Let B be a replay buffer storing the experience for training.
$\eta = \eta_{\text{init}}$.
Let $i \sim \text{Uniform}\{1 \dots M\}$ and $\psi = 1$ w.p. χ otherwise $\psi = 0$
for N epochs **do**
    for C steps **do**
        Observe s
        Choose $a = \arg\max_a Q_i(s,a) + \psi Q_1(s,a)$
        Observe reward r, next state s', end of episode t
        If t is terminal, $i \sim \text{Uniform}\{2 \dots M\}$ and $\psi = 1$ w.p. χ otherwise $\psi = 0$
        Store $< s, a, r, s', t >$ in buffer B
        Sample mini-batch B_{batch}
        Update $\{Q_k\}_{k=1}^K$ using equation (7.13)
        $Q \leftarrow Q +$ | violated constraints (7.14) in B_{batch}|
    end for
    Let $\rho = \frac{V}{C * \text{batch\_size}}$
    Update η by (7.15)
    Update target network
end for

---

where Q_k^T is the target network of the kth approximator. To get an unbiased performance evaluation, we decided to update $M - 1$ components of the ensemble with the update rule provided by BDQN. We make this choice in order to maintain diversity between the approximations of the ensemble as shown in (Osband et al., 2016). We use the remaining single component of the ensemble to approximate Q_E. Using the first component to approximate Q_E, we get for our new algorithm optimistic DQN (ODQN) the loss

$$\mathcal{L}_O(s,a) = \left(\eta^{-1} \log \frac{\sum_{k=2}^M e^{\eta \gamma \max_{a'} Q_k^T(s',a') + Q_1^T(s',a')}}{M} - \frac{\sum_{k=2}^M \gamma \max_{a'} Q_k^T(s',a')}{M} - Q_1(s,a) \right)^2$$

$$+ \sum_{k=2}^M \left(r + \gamma \max_{a'} Q_k^T(s',a') - Q_k(s,a) \right)^2. \tag{7.13}$$

The exploratory bonus represented by $Q_E = Q_1$ in the proposed OQL and OQDN algorithms is needed to guide exploration during learning. During evaluation, we use majority voting on the remaining $M - 1$ components $\{Q_k\}_{k=2}^M$. While we always select an optimistic policy in OQL during the training phase, in ODQN the neural network function approximator may have problems learning to approximate the optimal policy: if there are not enough unbiased samples the approximator may learn to model only the optimistic biased samples. Note that in the tabular case, this is not a problem since there is no Q-function approximation. In order to mitigate this problem, we introduce a hyper-parameter χ which denotes the probability to select an optimistic policy π_o in place of the unbiased one π_u. In this way, we can balance the number of unbiased and optimistic samples. Algorithm 3 shows the pseudocode of ODQN.

Automatic Hyper-parameter Adaptation. Recalling that the regularization coefficient η in the OBE is hard to tune, we want to focus our attention on Problem 2. Inspired by proximal policy optimization (PPO) (Schulman et al., 2017), we propose an heuristic to optimize η. One of the optimization techniques proposed in (Schulman et al., 2017) is to measure the "degree" of constraint violation and to update the Lagrangian multiplier accordingly. We have to adapt the technique to multiple constraints since the problem is defined for each state-action pair. The idea is to count the number of times the constraints have been violated and then update η. In more detail, suppose to have N state-action pairs and for each pair (s_i, a_i)

$$\sum_{m=1}^{M} b_m(s_i, a_i)(\log b_m(s_i, a_i) + \log M) \leq \iota_{\max}, \tag{7.14}$$

where $\iota_{\max}$ is defined in Problem 2, while $b_m(s_i, a_i)$ is defined by (7.3). We define ρ as the ratio of violated constraints. We update η according to the following rule

$$\eta_{T+1} = \frac{\eta_T}{(0.5 + 10\rho)}. \tag{7.15}$$

In ODQN, we decided to count the number of constraints violated every C time-steps (basically every update of the target network), using the samples of all the extracted mini-batches. See Algorithm 3 for further details.

Ensuring a Diversity in the Ensemble. As already discussed, it is important to maintain diversity in our ensemble, and this diversity should reflect the degree of uncertainty. For this reason, we should introduce a sort of prior distribution, as happens in the Bayesian framework. In the case of OQL, we observe that it is sufficient to randomly initialize each element of the ensemble, since diversity between estimates is a sufficient condition to obtain positive bonus. For ODQN, as is done in BDQN, we choose to maintain the diversity between approximation, by a random initialization of each component's parameters and by using the bootstrapping technique, so by adding a *mask* in the replay memory which is sampled by and use different data samples per regressor.

7.3. Empirical Analysis

In the experiments, we compare in the tabular Q-function case our new optimistic Q-learning method (OQL) with bootstrapped Q-learning method (BQL) - which is the tabular version of BDQN, the well-known state-of-the-art Q-learning (QL) (Watkins & Dayan, 1992), and Q-learning with optimistic initialization (OIQL) (Sutton & Barto, 2018) in the 50-Chain (Osband et al., 2016), Taxi (also known as Maze) (Dearden et al., 1998) and Frozen Lake (Brockman et al., 2016) environments. For neural-network based Q-functions, we compare our new optimistic deep Q-learning (ODQN) method with bootstrapped deep Q-learning (BDQN) and classical deep Q-learning (DQN) in the Taxi and Acrobot (Sutton, 1996) environments. The environments are chosen to cover different types of dynamics, have sparse rewards, and include both discrete and continuous states. For Acrobot and Frozen Lake, we used the implementation provided by OpenAI Gym (Brockman et al., 2016). First, we will discuss the environments in more detail, then we will provide some details on the initialization of the methods, and finally finish with an analysis of the results.

The **N-Chain** environment (Osband et al., 2016) requires a long sequence of non-rewarding actions to achieve the optimal reward. The MDP consists of a chain with N states $\{s_i\}_{i=1}^{N}$, and two actions that move the agent to state s_{i+1} or s_{i-1}. The agent always starts in state s_2. In state s_1 the agent observes a small reward $r(s_1) = 1/1000$, while in the Nth state the agent observes $r(s_N) = 1$. The reward function is zero elsewhere. The agent needs to explore until reaching state s_N even if state s_1 looks promising.

The **Taxi** environment, also known as Maze (Dearden et al., 1998), consists of a 8×8 grid-world where a taxi has to collect passengers and take them to the goal position where the only non-null reward is observed.

Figure 7.2.: Illustration of the environments, from the left to the right: N-Chain, Taxi, Frozen Lake and Acrobot.

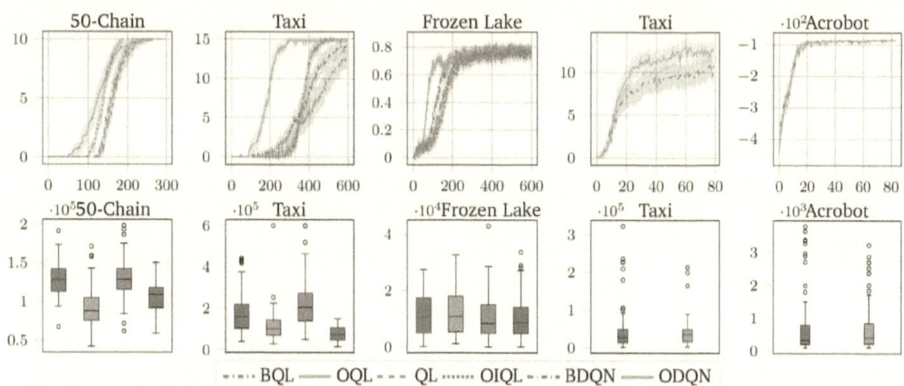

Figure 7.3.: The first row shows the average return for each tabular algorithm in the N-Chain, Taxi, and Frozen Lake environments and for each neural network based algorithm in the Taxi and Acrobot environments together with 95% confidence intervals. The second row shows the distribution over the number of time steps before observing the maximum reward in the MDP.

For 0, 1, 2 and 3 passengers collected, the reward is 0, 1, 3 and 15, respectively. The agent must explore to 1) discover that reaching the goal position with less than 3 passengers is not optimal, 2) to find the optimal path.

The **Frozen Lake** environment is a 8×8 grid-world, in which the agent has to reach a goal position without falling into some holes. The stochastic perturbation of the agent's movement makes the environment challenging.

The **Acrobot** environment was firstly proposed by (Sutton, 1996) and consists of two linked robotic arms that hang downward with only one actuated joint between the two arms. The agent swings the robotic arms until the end of the second link exceeds a certain height. At each step the agent perceives a negative reward of -1. The environment requires significant exploration to find a way to swing the robotic arms at defined height.

Initialization of the Q-functions. In the tabular setting, for optimistically initialized Q-learning (OIQL), we initialize the Q-function to 15 in Taxi and to 1 in N-Chain and Frozen Lake. For the other algorithms, we initialize $Q(s, a) \sim \mathcal{N}(\mu = 0, \sigma = 2)$, except Q_E of OQL is initialized to 0. In the taxi environment, for both ODQN and BDQN, we use a shared convolutional layer with multiple heads as described in (Osband et al., 2016). For the Acrobot, each component of the ensemble corresponds to a one-layer neural network. For ODQN, in both the environment, we initialize the output layer corresponding to Q_E to small values of the parameters, in order to obtain initially $Q_E \approx 0$.

Hyper-Parameter Tuning. For the tabular settings, we did not run any hyper-parameter optimization. With neural networks we performed a small grid search over the number of neurons of the network, and whether to use bootstrapping or not. More specifically, we selected hyper-parameters maximizing the mean return averaged over the whole learning curve, using 20 different seeds. In the plots; we compare ODQN and BDQN using the best hyper-parameter setup found for BDQN. In OQL we use $\eta = 10$ and for ODQN, we use $\chi = 0.25$ and $\iota_{\max} = 1$. For further details about the implementation of the algorithms, and the grid search, see Appendix D.

7.3.1. Results

Figure 7.3 summarizes the results obtained by averaging over 64 different seed the tabular algorithms, and 100 seeds BDQN and ODQN (more details, e.g. results for different hyper-parameter settings, can be found in Appendix D). OQL learns faster than the other tabular algorithms. In the Chain environment, and in Taxi, our algorithm OQL finds the highest reward faster than BQL or QL. On the other hand, also OIQL seems to find high rewards fast, as shown in the box-plots. However, OIQL requires more training epochs to escape the high initial optimistic values of the value function. In contrast, OQL finds high values fast in all the problems (50-Chain, Taxi, Frozen Lake) using the optimistic Bellman equation while converging to a near-optimal solution as suggested by our convergence proofs. The algorithm finds the global optimum, which shows that the amount of exploration decreases, as the increasing amount of samples lower the epistemic uncertainty. The introduction of function approximation in ODQN results effective as it outperforms BDQN in the Taxi environment. In the Acrobot environment, the learning curves of BDQN and ODQN are nearly identical, possibly, due to the simplicity of the environment.

7.4. Epilogue

The main contribution of our work is the introduction of the optimistic Bellman equation (OBE) which provides an optimistic estimate of the value function over uncertainty. Our approach can be viewed as a principled IM technique where the agent is intrinsically rewarded by uncertainty and which, similar to approximated Bayesian methods, estimates the an ensemble. We propose two algorithms: OQL for the tabular case and ODQN for the neural network case. Given the usual assumptions on the learning rate and state visits, we show that OQL convergences to the optimal policy, analyze the implicitly defined exploration bonus in OQL and show the relationship to intrinsic motivation based approaches. In empirical evaluations on a variety of tasks where exploration is crucial, OQL and OQDN show to perform better than the comparison methods.

8. Conclusion

To provide more efficient and reliable robot-learning, we examined different techniques, spanning from dimensionality reduction, robotics movement representation, episodic and step-based off-policy reinforcement learning, statistical guarantees on kernel regression, and exploration in large Markov decision processes. For each of the examined technique, we provide theoretical motivations and analysis, and an empirical evaluation conducted in simulation, and, in some cases, on real-robots.

8.1. Summary of the Contributions

This book empirically proves that an efficient representation of robotic movements united with an off-policy gradient estimation solves high-dimensional tasks (such as closing a drawer placed in different positions) with few-hundreds demonstrations and subsequent interactions. Real-robotic experiments confirmed the results. Chapter 5 shows that a novel non-parametric off-policy gradient technique that trades bias and variance adequately offers a precedently unseen sample efficiency, for example, in the mountain car tasks, where the algorithm elaborates an optimal policy using only two sub-optimal demonstrations. Chapter 6 details a novel bound on the bias on a particular kernel regression technique, showing that simple models are can provide necessary statistical guarantees, which we believe are needed in the application of machine learning to real-world scenarios. Last but not least, Chapter 7 offers a new far-sighted exploration technique, proving on classical tasks a better sample efficiency when compared to state-of-the-art techniques.

We summarize our take-away messages.

1. Choosing the right model is essential to provide safe and efficient policy improvement. In our case, utilizing dimensionality reduction combined with movement primitives allowed us to improve the movements in a convenient latent space. Similarly, we showed that in off-policy improvement, non-parametric estimates overcome issues presents in other models.

2. Simple statistical models, such as mixtures of probabilistic component analyzers and Nadaraya-Watson kernel regression, offer the possibility of achieving convenient mathematics (i.e., conditioning and KL-bounds in close forms in the case of a mixture of principal component analyzers and closed-form solution of the value function with Nadaraya-Watson kernel regression). Furthermore, thanks to their simplicity, simple estimators allow strong statistical guarantees.

3. Off-policy estimates are essential to safe and efficient reinforcement learning. However, given their inherent difficulty, they must be performed carefully, with special care of their bias-variance trade-off.

8.1.1. The Impact of Dimensionality Reduction for Policy Improvement

Dimensionality reduction is at the core of Chapter 3 and 4. We argued and proved that for robotic manipulation tasks, a dimensionality reduction on the movement parameters is more effective than the more classic reduction in the joint space. This result is explained by the fact that the degrees of freedom are relatively small in robotic manipulation, and they offer low redundancies, while tasks often share more similarities and redundancies. In Chapter 4, we suggested using a mixture of principal components analyzers, which trades mathematical tractability with powerful expressibility, allowing the detection of multi modalities in the dataset. We stress that dimensionality reduction offers a convenient tool to represent the movements in a convenient latent space, allowing the subsequent reinforcement learning to safely explore and be more efficient, given the lower number of parameters.

8.1.2. Statistical Models and Guarantees

As previously discussed, in this book, we used movement primitives, which are linear models, a mixture of probabilistic principal component analyzers, which can be seen as a particular case of Gaussian mixture models, and in Chapter 5, we used Nadaraya-Watson kernel regression. The choice of such statistical methods might seem outdated in an era where deep models are leading the research. Deep models have shown tremendous performances in high-dimensional tasks, such as image-classification, text-to-speech, automated text generations. However, these tasks' dimensionality is of much larger dimensionality of the tasks analyzed in this book. Nevertheless, some of our considered tasks deal with 200 dimensions, which is certainly not a small amount. In such a situation, we showed that a careful choice of the model still performs well under this condition (in Section 2.1 we showed the mixture of probabilistic principal component analyzers working on the MNIST dataset, and in Chapter 4 working on the 200 dimensions of a robotic task), allowing mathematical tractability (statistical conditioning, KL-divergence, and value estimation in closed form). Furthermore, in the case of Nadaraya-Watson regression, we were able to furnish a hard upper bound on the bias (Chapter 6), which allowed us to upper bound the bias of the value estimation in Chapter 5. To conclude, while deep models are undoubtedly useful in very high dimensional problems, for robotic tasks, uncomplicated but still expressive statistical models help reaching a better sample efficiency and safety as they enjoy favorable mathematical properties that are simpler to analyze.

8.1.3. Bias and Variance in Off-Policy Gradient Estimation

Off-policy methods hold the promise of more sample efficiency and better safety as they (i) allow sample reuse and (ii) allow the extraction of the optimal policy from safe interaction (i.e., presented by an expert). However, off-policy estimates are hard to obtain. In the literature, off-policy estimates can be obtained via semi-gradient approaches, which suffer from high bias or importance-sampling approaches that suffer from high variance. In Chapter 4, we argued that a self-normalized importance sampling united with a full gradient expansion of the normalizer offers a good trade-off and reliable gradient in the simple case of episodic reinforcement learning. In the more complex case of step-based approaches, however, the variance of importance sampling grows multiplicatively. A closed-form solution of a non-parametric Bellman equation allows a better trade-off, tunable via the kernel's bandwidth.

8.2. Future Work

Our research successfully proposed novel techniques to enhance learning on real robotic systems. However, many research questions remain open. In Chapters 3 and 4, we focused on tasks solvable with one movement, and we used episodic reinforcement learning. In a more realistic scenario, complex tasks need more than one step to be solved, and step-based reinforcement learning in therefore needed. We treated step-based reinforcement learning in Chapters 5 and 7 to provide sample efficient off-policy improvement and far-sighted exploration. However, despite the promising results, the method proposed in Chapter 5, due to its nonparametric nature, is limited to low-dimensional tasks. On the other hand, exploration in Chapter 7 is limited to a discrete set of actions, hindering its application to real robotics. However, efficient step-based reinforcement learning solves only one part of the many problems present in a real industrial environment. In this book, we consider safety by projecting the policy's parameters to a latent space. In literature, however, the problem is more articulated (Garcia & Fernández, 2015). In an industrial robotic scenario, we often need hard constraints on joint velocities, collision avoidance, to mention a few. A safe representation of the policy is often not enough to solve these issues. We detail the future work by first building on the methods proposed in this book, and we terminate the chapter with a more far-sighted view that comprehends the problem of learning for industrial robotics in a larger picture.

Bridging Step-Based Reinforcement Learning and Latent Space Movement Primitives. The efficacy of the mixture of probabilistic principal component analyzers when applied to relatively high dimensional tasks and its efficacy in solving complex robotic tasks suggests that it can also be employed in the more complex step-based scenario presented in Chapter 5. In detail, the nonparametric Bellman equation is derived as a conditional density estimation. Density estimation can also be achieved with mixtures of probabilistic principal component analyzers, allowing a more powerful formulation of the Bellman equation solvable in closed-form. This newer Bellman equation allows its usage in greater dimensional problem and, in contrast to nonparametric methods, scales with the number of samples. Allowing the mixture of principal component analysis to solve the Bellman recursion providing a full gradient estimate provides an efficient off-policy improvement, while the latent projection should keep a convenient and safe representation of the movement.

Optimistic Bellman Equation with Last-Layer Bayesian Linear Regression. The results provided in Chapter 7 are promising, but they are limited to low-dimensional tasks. On the one hand, ensemble models provide a simple estimation of the epistemic uncertainty, but on the other hand, the estimates are affected by high variance. Bayesian linear regression enjoys a more robust theoretical foundation and often delivers a better estimate. Last-layer Bayesian linear regression holds the theoretical foundations and provides an automatic feature extraction that works in high dimensions (Azizzadenesheli et al., 2018). For this reason, we believe that its usage can substantially empower our method.

Exploitation of the Model to Include Constrained Optimization. The nonparametric off-policy gradient described in Chapter 5 builds internally a model which can be considered a transition matrix. This model appears in any closed-form solution of a Bellman equation. For example, the closed-form of a Bellman equation with linear parametrization builds a similar matrix that encodes transitions between features. The usage of this matrix can go beyond the gradient computation. We can use this matrix to build a constrained problem to ensure safe behavior. In Chapter 7, we have seen the importance of distinguishing the epistemic uncertainty from the aleatoric one. In Chapters 4 and 5, the probabilistic models take in account of the sole aleatoric uncertainty. To guarantee a safe and efficient exploration, one needs to distinguish the epistemic and the aleatoric uncertainty carefully (Clements et al., 2019). While the epistemic uncertainty can be corrected by acquiring more samples and boosting the model (to lower the model bias), the aleatoric source of uncertainty can only be lowered by providing a better policy that avoids stochastic transitions.

Inverse Reinforcement Learning and Human in the Loop.   The definition of the reward signal remains an obstacle to fully automated learning as it requires a human expert. While exploration partially solves the problem, as it allows more straightforward reward definitions, inverse reinforcement learning holds the promise of extracting the reward from human demonstrations. This argument supports the possibility to augment the imitation-learning phase by introducing the reward signal's automatic extraction by inverse reinforcement learning. However, this pipeline limits human intervention only to the beginning of the learning phase. In human beings, the human demonstrator (or teacher) continuously interacts with the student providing demonstrations and advice. This continuous interaction ensures more stable and efficient learning. In contrast to the classic pipeline, recent advances in the literature (Hadfield-Menell et al., 2016, 2017) suggest the formalization of the reinforcement learning problem as a multiagent instance where the human cooperate with the robot, and the robot must optimize the human's internal reward. We believe that this framing of the problem is more appealing than the classic formulation. For this reason, as future work, we will study how to develop a new off-policy improvement built on this new framework.

Appendices

A. Dimensionality Reduction of Movement Primitives in Parameter Space

In this appendix, we show a more extended empirical analysis of the correlations, and of the dimensionality techniques applied to different datasets.

A.1. Correlations in the Datasets

Figure A.1 depicts the correlation between joints, and correlation between parameters. It is possible to observe that the parameter space captures more redundancies.

A.2. Dimensionality Reduction

Figure A.2 depicts the accuracy of reconstruction of the dimensionality reduction applied to different datasets. It is possible to observe that the dimensionality reduction applied in the parameter space is always convenient, both in terms of accuracy and of number of parameters needed.

A.3. Imitation Learning

Figure A.3 depicts the accuracy in reconstructing a movement given a specific context. Although the reconstruction error achieved in parameter space is not sensibly better than the one in joint space, it requires a significant number of parameter less.

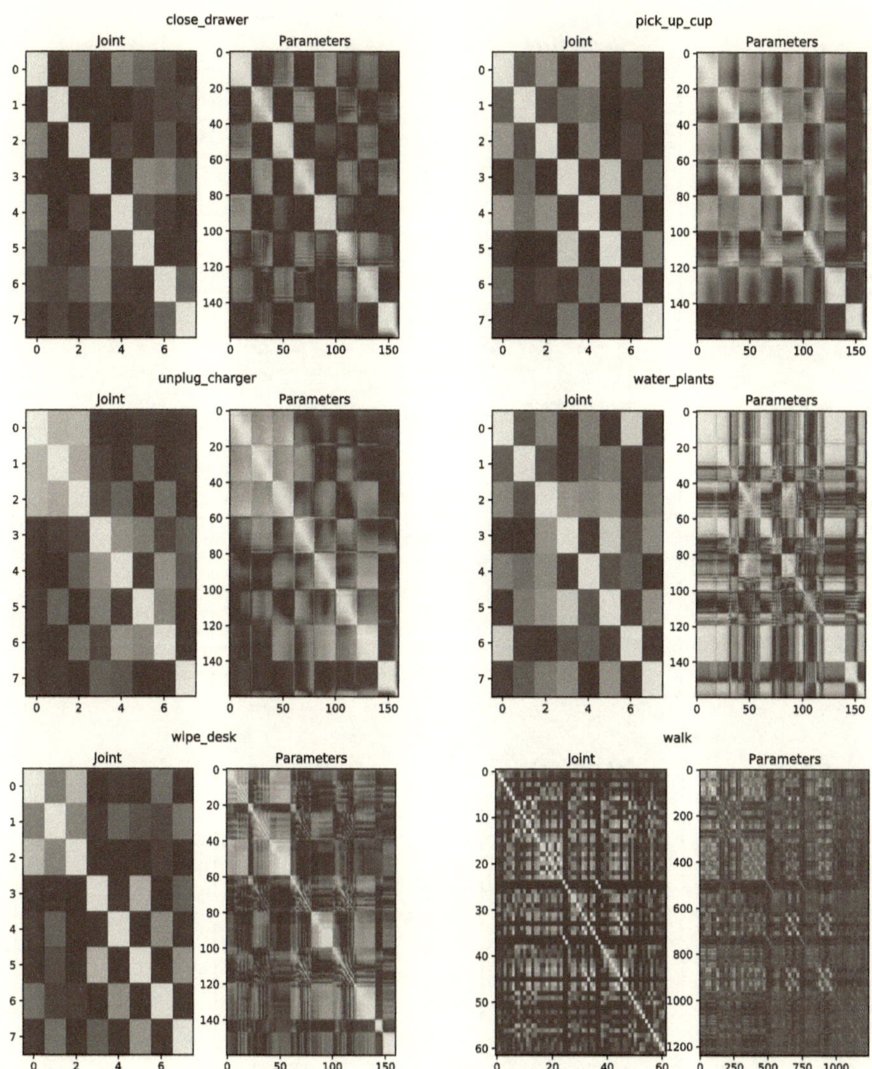

Figure A.1.: Correlation in joint and parameter space in different datasets.

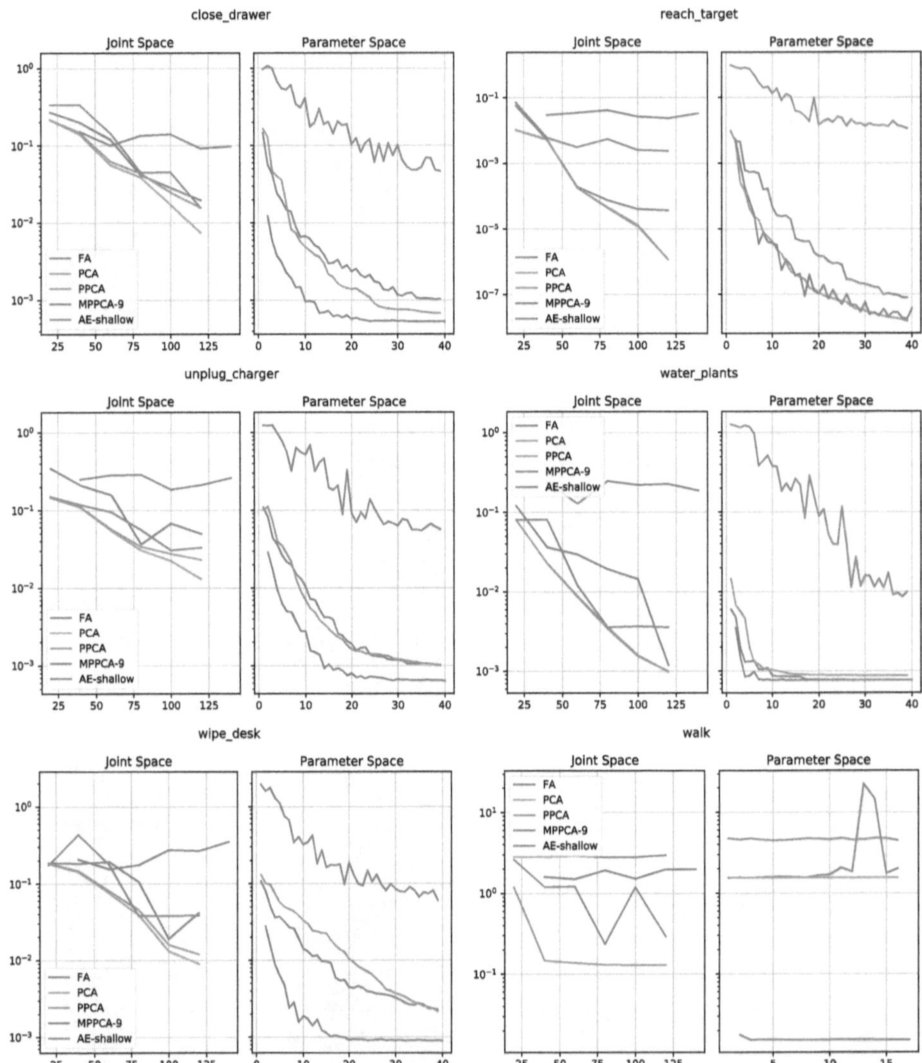

Figure A.2.: Dimensionality reduction applied to the movement primitives on different datasets.

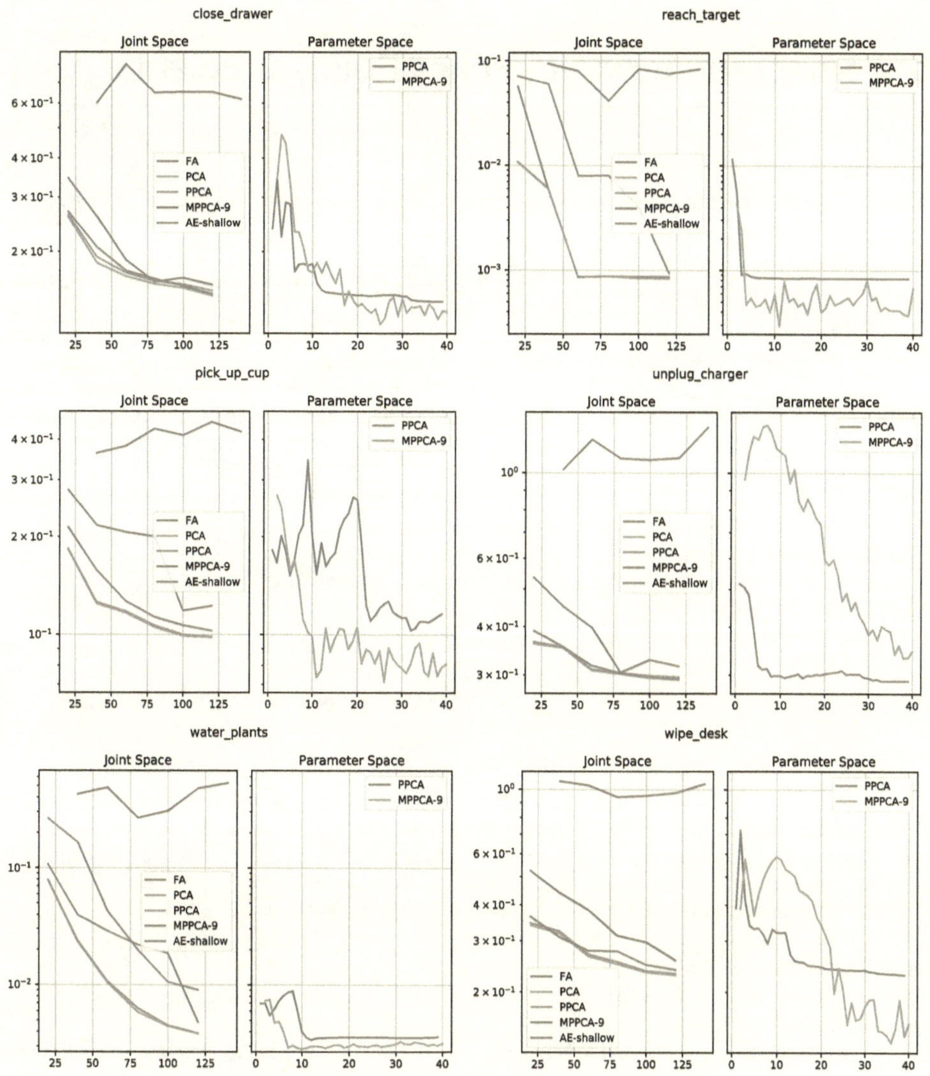

Figure A.3.: Imitation learning on different datasets.

B. A Nonparametric Off-Policy Policy Gradient

In this appendix we show the proofs of Theorem 1, Theorem 3 and we detail experimental settings.

B.1. Existence and Uniqueness of the Fixed Point of the Nonparametric Bellman Equation

In the following, we discuss the existence and uniqueness of the solution of the nonparametric Bellman equation (NPBE). To do so, we will first notice that the solution must be a linear combination of the responsability vector $\varepsilon_\pi(\mathbf{s})$, and therefore any solution of the NPBE must be bounded. We will also show that such a bound is $\pm R_{\max}/(1 - \gamma_c)$. We will use this bound to show that the solution must be unique. Subsequently, we will prove Theorem 1.

Proposition 1. Space of the Solution
The solution of the NBPE is $\varepsilon_\pi^\mathsf{T}(\mathbf{s})\mathrm{q}$ where $\mathrm{q} \in \mathbb{R}^n$, where n is the number of support points.
Informal Proof: The NPBE is

$$\hat{V}_\pi(\mathbf{s}) = \varepsilon_\pi^\mathsf{T}(\mathbf{s}) \left(\mathbf{r} + \int_\mathcal{S} \phi_\gamma(\mathbf{s}')\hat{V}_\pi(\mathbf{s}')\,\mathrm{d}\mathbf{s}' \right) \tag{B.1}$$

Both $\mathbf{r}$ and $\int_\mathcal{S} \phi_\gamma(\mathbf{s}')\hat{V}_\pi(\mathbf{s}')\,\mathrm{d}\mathbf{s}'$ are constant w.r.t. $\mathbf{s}$, therefore $\hat{V}_\pi(\mathbf{s})$ must be a linear respect to $\varepsilon_\pi(\mathbf{s})$.

A first consequence of Proposition 1 is that, since $\varepsilon_\pi(\mathbf{s})$ is a vector of kernels, $\hat{V}_\pi(\mathbf{s})$ is bounded.

In Theorem 1 we state that the solution of the NPBE is $\hat{V}_\pi^*(\mathbf{s}) = \varepsilon_\pi^\mathsf{T}(\mathbf{s})\Lambda_\pi^{-1}\mathbf{r}$. It is trivial to show that such a solution is a valid solution of the NPBE; however, the uniqueness of such a solution is non-trivial. In order to prove it, we will before show that if a solution to the NPBE exists, then the solution must be bounded by $[-R_{\max}/(1 - \gamma_c), R_{\max}/(1 - \gamma_c)]$, where $R_{\max} = \max_i |r_i|$. Note that this nice property is not common for other policy-evaluation algorithms (e.g., Neural Fitted Q-Iterations Riedmiller (2005)).

Proposition 2. Bound of NPBE
If $V_\pi : \mathcal{S} \to \mathbb{R}$ is a solution to the NPBE, then $|V_\pi(\mathbf{s})| \leq R_{\max}/(1 - \gamma_c)$.

Proof. Suppose, by contradiction, that a function $f(\mathbf{s})$ is the solution of a NPBE, and that $\exists \mathbf{z} \in \mathcal{S} : f(\mathbf{z}) = R_{\max}/(1 - \gamma_c) + \epsilon$ where $\epsilon > 0$. Since the solution of the NPBE must be bounded, we can further assume without any loss of generality that $f(\mathbf{s}) \leq f(\mathbf{z})$. Then,

$$\frac{R_{\max}}{1 - \gamma_c} + \epsilon = \varepsilon_\pi^\mathsf{T}(\mathbf{z})\mathbf{r} + \varepsilon_\pi^\mathsf{T}(\mathbf{z}) \int_\mathcal{S} \phi_\gamma(\mathbf{s}')f(\mathbf{s}')\,\mathrm{d}\mathbf{s}'.$$

Since by assumption the previous equation must be fulfilled, and $\varepsilon_\pi(\mathbf{z})$ is a stochastic vector, $|\varepsilon_\pi^\mathsf{T}(\mathbf{z})\mathbf{r}| \leq R_{\max}$, we have

$$\left| \frac{R_{\max}}{1-\gamma_c} + \epsilon - \varepsilon_\pi^\mathsf{T}(\mathbf{z}) \int_S \phi_\gamma(\mathbf{s}')f(\mathbf{s}')\,\mathrm{d}\mathbf{s}' \right| \leq R_{\max}. \tag{B.2}$$

However, noticing also that $0 \leq \phi_i^{\pi,\gamma}(\mathbf{s}) \leq \gamma_c$, we have

$$
\begin{aligned}
\frac{R_{\max}}{1-\gamma_c} + \epsilon - \varepsilon_\pi^\mathsf{T}(\mathbf{z}) \int_S \phi_\gamma(\mathbf{s}')f(\mathbf{s}')\,\mathrm{d}\mathbf{s}' \;&\geq\; \frac{R_{\max}}{1-\gamma_c} + \epsilon - \gamma_c \max_{\mathbf{s}} f(\mathbf{s}) \\
&\geq\; \frac{R_{\max}}{1-\gamma_c} + \epsilon - \gamma_c \frac{R_{\max}}{1-\gamma_c} \\
&=\; R_{\max} + \epsilon,
\end{aligned}
$$

which is in contradiction with Equation B.2. A completely symmetric proof can be derived assuming by contradiction that $\exists \mathbf{z} \in S : f(\mathbf{z}) = -R_{\max}/(1-\gamma_c) - \epsilon$ and $f(\mathbf{s}) \geq f(\mathbf{z})$. $\qquad\square$

Proposition 3. *If $\mathbf{r}$ is bounded by R_{max} and if $f^* : S \to \mathbb{R}$ satisfies the NPBE, then there is no other function $f : S \to \mathbb{R}$ for which $\exists \mathbf{z} \in S$ and $|f^*(\mathbf{z}) - f(\mathbf{z})| > 0$.*

Proof. Suppose, by contradiction, that exists a function $g : S \to \mathbb{R}$ such that $f^*(\mathbf{s}) + g(\mathbf{s})$ satisfies Equation B.1. Furthermore assume that $\exists \mathbf{z} : g(\mathbf{z}) \neq 0$. Note that, since $f : S \to \mathbb{R}$ is a solution of the NPBE, then

$$\int_S \varepsilon_\pi^\mathsf{T}(\mathbf{s})\phi_\gamma(\mathbf{s}')f^*(\mathbf{s}')\,\mathrm{d}\mathbf{s}' \in \mathbb{R}, \tag{B.3}$$

and similarly

$$\int_S \varepsilon_\pi^\mathsf{T}(\mathbf{s})\phi_\gamma(\mathbf{s}')\left(f^*(\mathbf{s}') + g(\mathbf{s}')\right)\,\mathrm{d}\mathbf{s}' \in \mathbb{R}. \tag{B.4}$$

The existence of the integrals in Equations B.3 and B.4 implies

$$\int_S \varepsilon_\pi^\mathsf{T}(\mathbf{s})\phi_\gamma(\mathbf{s}')g(\mathbf{s}')\,\mathrm{d}\mathbf{s}' \in \mathbb{R}. \tag{B.5}$$

Note that

$$
\begin{aligned}
|g(\mathbf{s})| \;&=\; |f^*(\mathbf{s}) - (f^*(\mathbf{s}) + g(\mathbf{s}))| \\
&=\; \left| f^*(\mathbf{s}) - \varepsilon_\pi^\mathsf{T}(\mathbf{s})\left(\mathbf{r} + \int_S \phi_\gamma(\mathbf{s}')(f^*(\mathbf{s}') + g(\mathbf{s}'))\,\mathrm{d}\mathbf{s}'\right) \right| \\
&=\; \left| \varepsilon_\pi^\mathsf{T}(\mathbf{s})\left(\mathbf{r} + \int_S \phi_\gamma(\mathbf{s}')f^*(\mathbf{s}')\,\mathrm{d}\mathbf{s}'\right) - \varepsilon_\pi^\mathsf{T}(\mathbf{s})\left(\mathbf{r} + \int_S \phi_\gamma(\mathbf{s}')(f^*(\mathbf{s}') + g(\mathbf{s}'))\,\mathrm{d}\mathbf{s}'\right) \right| \\
&=\; \left| \varepsilon_\pi^\mathsf{T}(\mathbf{s}) \int_S \phi_\gamma(\mathbf{s}')g(\mathbf{s}')\,\mathrm{d}\mathbf{s}' \right| \\
&=\; \left| \varepsilon_\pi^\mathsf{T}(\mathbf{s}) \int_S \phi_\gamma(\mathbf{s}')g(\mathbf{s}')\,\mathrm{d}\mathbf{s}' \right|.
\end{aligned}
$$

Using Jensen's inequality

$$|g(\mathbf{s})| \;\leq\; \varepsilon_\pi^\mathsf{T}(\mathbf{s}) \int_S \phi_\gamma(\mathbf{s}')|g(\mathbf{s}')|\,\mathrm{d}\mathbf{s}'.$$

Since both f^* and $f + g$ are bounded by $\frac{R_{\max}}{1-\gamma_c}$, then $|g(\mathbf{s})| \leq A = \frac{2R_{\max}}{1-\gamma_c}$, then

$$|g(\mathbf{s})| \;\leq\; \varepsilon_\pi^\mathsf{T}(\mathbf{s}) \int_{\mathcal{S}} \phi_\gamma(\mathbf{s}')|g(\mathbf{s}')|\,\mathrm{d}\mathbf{s}' \tag{B.6}$$

$$\leq\; A\varepsilon_\pi^\mathsf{T}(\mathbf{s}) \int_{\mathcal{S}} \phi_\gamma(\mathbf{s}')\,\mathrm{d}\mathbf{s}'$$

$$\leq\; \gamma_c A.$$

We can iterate this reasoning now posing $|g(\mathbf{s})| \leq \gamma_c A$, and eventually we notice that $|g(\mathbf{s})| \leq 0$, which is in contradiction with the assumption made. $\qquad\square$

Proof of Theorem 1

Proof. Saying that $\hat{V}_\pi^*$ is a solution for Equation B.1 is equivalent to say

$$\hat{V}_\pi^*(\mathbf{s}) - \varepsilon_\pi^\mathsf{T}(\mathbf{s})\left(\mathbf{r} + \gamma \int_{\mathcal{S}} \phi_\gamma(\mathbf{s}')\hat{V}_\pi^*(\mathbf{s}')\,\mathrm{d}\mathbf{s}'\right) = 0 \qquad \forall \mathbf{s} \in \mathcal{S}.$$

We can verify that by simple algebraic manipulation

$$\hat{V}_\pi^*(\mathbf{s}) - \varepsilon_\pi^\mathsf{T}(\mathbf{s})\left(\mathbf{r} + \int_{\mathcal{S}} \phi_\gamma(\mathbf{s}')\hat{V}_\pi^*(\mathbf{s}')\,\mathrm{d}\mathbf{s}'\right)$$

$$= \; \varepsilon_\pi^\mathsf{T}(\mathbf{s})\Lambda_\pi^{-1}\mathbf{r} - \varepsilon_\pi^\mathsf{T}(\mathbf{s})\left(\mathbf{r} + \int_{\mathcal{S}} \phi_\gamma(\mathbf{s}')\varepsilon_\pi^\mathsf{T}(\mathbf{s}')\Lambda_\pi^{-1}\mathbf{r}\,\mathrm{d}\mathbf{s}'\right)$$

$$= \; \varepsilon_\pi^\mathsf{T}(\mathbf{s})\left(\Lambda_\pi^{-1}\mathbf{r} - \mathbf{r} - \int_{\mathcal{S}} \phi_\gamma(\mathbf{s}')\varepsilon_\pi^\mathsf{T}(\mathbf{s}')\Lambda_\pi^{-1}\mathbf{r}\,\mathrm{d}\mathbf{s}'\right)$$

$$= \; \varepsilon_\pi^\mathsf{T}(\mathbf{s})\left(\left(I - \int_{\mathcal{S}} \phi_\gamma(\mathbf{s}')\varepsilon_\pi^\mathsf{T}(\mathbf{s}')\,\mathrm{d}\mathbf{s}'\right)\Lambda_\pi^{-1}\mathbf{r} - \mathbf{r}\right)$$

$$= \; \varepsilon_\pi^\mathsf{T}(\mathbf{s})\left(\Lambda_\pi\Lambda_\pi^{-1}\mathbf{r} - \mathbf{r}\right)$$

$$= \; 0. \tag{B.7}$$

Since equation B.1 has (at least) one solution, Proposition 3 guarantees that the solution $(\hat{V}_\pi^*)$ is unique. $\quad\square$

B.2. Bias of the Nonparametric Bellman Equation

In this section, we want to prove Theorem 3. To do so, we introduce the infinite-samples extension of the NPBE.

Proposition 4. *Let us suppose to have a dataset of infinite samples, and in particular one sample for each state-action pair of the state-action space. In the limit of infinite samples the NPBE defined in Definition 2 with a data-set $\lim_{n\to\infty} D_n$ collected under distribution β on the state-action space and MDP $\mathcal{M}$ converges to*

$$V_D(\mathbf{s}) = \lim_{n\to\infty} \int_{\mathcal{A}} \frac{\sum_{i=1}^n \psi_i(\mathbf{s})\varphi_i(\mathbf{a})\left(r_i + \gamma_i \int_{\mathcal{S}} \phi_i(\mathbf{s}')\hat{V}_\pi(\mathbf{s}')\,\mathrm{d}\mathbf{s}\right)}{\sum_{j=1}^n \psi_j(\mathbf{s})\varphi_j(\mathbf{a})} \pi(\mathbf{a}|\mathbf{s})\,\mathrm{d}\mathbf{a}$$

$$= \int_{\mathcal{A}} \frac{\lim_{n\to\infty} \frac{1}{n}\sum_{i=1}^{n} \psi_i(\mathbf{s})\varphi_i(\mathbf{a})\left(r_i + \gamma_i \int_{\mathcal{S}} \phi_i(\mathbf{s}')\hat{V}_\pi(\mathbf{s}')\,\mathrm{d}\mathbf{s} \right)}{\lim_{n\to\infty} \frac{1}{n}\sum_{j=1}^{n} \psi_j(\mathbf{s})\varphi_j(\mathbf{a})} \pi(\mathbf{a}|\mathbf{s})\,\mathrm{d}\mathbf{a}$$

$$= \int_{\mathcal{A}} \frac{\int_{\mathcal{S}\times\mathcal{A}} \psi(\mathbf{s},\mathbf{z})\varphi(\mathbf{a},\mathbf{b})\left(R(\mathbf{z},\mathbf{b}) + \int_{\mathcal{S}} \phi(\mathbf{s}',\mathbf{z}')p_\gamma(\mathbf{z}'|\mathbf{b},\mathbf{z})\hat{V}_\pi(\mathbf{s}')\,\mathrm{d}\mathbf{s} \right)\beta(\mathbf{z},\mathbf{b})\,\mathrm{d}\mathbf{z}\,\mathrm{d}\mathbf{b}}{\int_{\mathcal{S}\times\mathcal{A}} \psi(\mathbf{s},\mathbf{z})\varphi(\mathbf{a},\mathbf{b})\beta(\mathbf{z},\mathbf{b})\,\mathrm{d}\mathbf{z}\,\mathrm{d}\mathbf{b}} \pi(\mathbf{a}|\mathbf{s})\,\mathrm{d}\mathbf{a}. \quad \text{(B.8)}$$

If we impose the process generating samples to be non-degenerate distribution, and $-R_{\max} \leq R \leq R_{\max}$, we see that Propositions 1-3 remain valid. Furthermore, from Equation B.8 we are able to infer that $\mathbb{E}_D[V_D(\mathbf{s})] = V_D(\mathbf{s})$ (since D is an infinite dataset, it does not matter if we re-sample it, the resulting value-function will be always the same).

Proof of Theorem 3. To keep the notation uncluttered, let us introduce

$$\varepsilon(\mathbf{s},\mathbf{a},\mathbf{z},\mathbf{b}) = \frac{\psi(\mathbf{s},\mathbf{z})\varphi(\mathbf{a},\mathbf{b})\beta(\mathbf{z},\mathbf{b})}{\int_{\mathcal{S}\times\mathcal{A}} \psi(\mathbf{s},\mathbf{z})\varphi(\mathbf{a},\mathbf{b})\beta(\mathbf{z},\mathbf{b})\,\mathrm{d}\mathbf{z}\,\mathrm{d}\mathbf{b}}. \quad \text{(B.9)}$$

We want to bound

$$\begin{aligned}
\overline{V}(\mathbf{s}) - V^*(\mathbf{s}) &= \int_{\mathcal{A}} \Bigg(\int \varepsilon(\mathbf{s},\mathbf{a},\mathbf{z},\mathbf{b})R(\mathbf{z},\mathbf{b})\,\mathrm{d}\mathbf{z}\,\mathrm{d}\mathbf{b} \\
&\quad + \int_{\mathcal{S}\times\mathcal{A}} \varepsilon(\mathbf{s},\mathbf{a},\mathbf{z},\mathbf{b}) \int_{\mathcal{S}} \phi_i(\mathbf{s}',\mathbf{z}')p_\gamma(\mathbf{z}'|\mathbf{b},\mathbf{z})\overline{V}(\mathbf{s}')\,\mathrm{d}\mathbf{s}'\,\mathrm{d}\mathbf{z}\,\mathrm{d}\mathbf{b} \\
&\quad - R(\mathbf{s},\mathbf{a}) - \int_{\mathcal{S}} V^*(\mathbf{s}')p_\gamma(\mathbf{s}'|\mathbf{s},\mathbf{a})\,\mathrm{d}\mathbf{s}' \Bigg) \pi(\mathbf{a}|\mathbf{s})\,\mathrm{d}\mathbf{a}
\end{aligned}$$

$$\begin{aligned}
\implies |\overline{V}(\mathbf{s}) - V^*(\mathbf{s})| &\leq \max_{\mathbf{a}} \left| \int \varepsilon(\mathbf{s},\mathbf{a},\mathbf{z},\mathbf{b})\left(R(\mathbf{z},\mathbf{b}) - R(\mathbf{s},\mathbf{a}) \right)\mathrm{d}\mathbf{z}\,\mathrm{d}\mathbf{b} \right| && \text{(B.10)} \\
&\quad + \max_{\mathbf{a}} \left| \int_{\mathcal{S}} p_\gamma(\mathbf{z}'|\mathbf{b},\mathbf{z})\left(\phi_i(\mathbf{s}',\mathbf{z}')\overline{V}(\mathbf{s}') - V^*(\mathbf{s}') \right)\mathrm{d}\mathbf{z}\,\mathrm{d}\mathbf{b} \right|
\end{aligned}$$

$$\begin{aligned}
&\leq \max_{\mathbf{a}} \underbrace{\left| \int \varepsilon(\mathbf{s},\mathbf{a},\mathbf{z},\mathbf{b})\left(R(\mathbf{z},\mathbf{b}) - R(\mathbf{s},\mathbf{a}) \right)\mathrm{d}\mathbf{z}\,\mathrm{d}\mathbf{b} \right|}_{A} && \text{(B.11)} \\
&\quad + \gamma_c \max_{\mathbf{a}} \underbrace{\left| \int_{\mathcal{S}\times\mathcal{A}} \varepsilon(\mathbf{s},\mathbf{a},\mathbf{z},\mathbf{b}) \int_{\mathcal{S}} p(\mathbf{z}'|\mathbf{b},\mathbf{z})\left(\phi_i(\mathbf{s}',\mathbf{z}')\overline{V}(\mathbf{s}') - V^*(\mathbf{s}') \right)\mathrm{d}\mathbf{s}'\,\mathrm{d}\mathbf{z}\,\mathrm{d}\mathbf{b} \right|}_{B}
\end{aligned}$$

$$= A_{\text{Bias}} + \gamma_c B_{\text{Bias}} \quad \text{(B.12)}$$

Term A is the bias of Nadaraya-Watson kernel regression, as it is possible to observe in Tosatto et al. (2020a), therefore Theorem 2 applies

$$A_{\text{Bias}} = \frac{L_R \sum_{k=1}^{d} \mathbf{h}_k \left(\prod_{i\neq k}^{d} e^{\frac{L_\beta^2 \mathbf{h}_i^2}{2}} \left(1 + \mathrm{erf}\left(\frac{\mathbf{h}_i L_\beta}{\sqrt{2}} \right) \right) \right)\left(\frac{1}{\sqrt{2\pi}} + L_\beta \mathbf{h}_k \frac{e^{\frac{L_\beta^2 \mathbf{h}_k^2}{2}}}{2} \left(1 + \mathrm{erf}\left(\frac{\mathbf{h}_k L_\beta}{\sqrt{2}} \right) \right) \right)}{\prod_{i=1}^{d} e^{\frac{L_\beta^2 \mathbf{h}_i^2}{2}} \left(1 - \mathrm{erf}\left(\frac{\mathbf{h}_i L_\beta}{\sqrt{2}} \right) \right)},$$

where $\mathbf{h} = [\mathbf{h}_\psi, \mathbf{h}_\varphi]$ and $d = d_s + d_a$.

$$
\begin{aligned}
B_{\text{bias}} &\leq \gamma_c \max_{\mathbf{a}} \left| \frac{\int_{\mathcal{S}\times\mathcal{A}} \psi(\mathbf{s},\mathbf{z})\varphi(\mathbf{a},\mathbf{b})\left(\int_{\mathcal{S}\times\mathcal{S}} \overline{V}(\mathbf{z}')\phi(\mathbf{z}',\mathbf{s}')p(\mathbf{s}'|\mathbf{s},\mathbf{a})\,\mathrm{d}\mathbf{s}'\,\mathrm{d}\mathbf{z}' - \int_{\mathcal{S}} V^*(\mathbf{s}')p(\mathbf{s}'|\mathbf{s},\mathbf{a})\,\mathrm{d}\mathbf{s}'\right)\beta(\mathbf{z},\mathbf{b})\,\mathrm{d}\mathbf{z}\,\mathrm{d}\mathbf{b}}{\int_{\mathcal{S},\mathcal{A}} \psi(\mathbf{s},\mathbf{z})\varphi(\mathbf{a},\mathbf{b})\beta(\mathbf{z},\mathbf{b})\,\mathrm{d}\mathbf{z}\,\mathrm{d}\mathbf{b}} \right| \\[2mm]
&= \gamma_c \max_{\mathbf{a}} \left| \frac{\int_{\mathcal{S}\times\mathcal{A}} \psi(\mathbf{s},\mathbf{z})\varphi(\mathbf{a},\mathbf{b})\left(\int_{\mathcal{S}}\int_{\mathcal{S}} \left(\overline{V}(\mathbf{z}')\phi(\mathbf{z}',\mathbf{s}') - V^*(\mathbf{s}')\right)p(\mathbf{s}'|\mathbf{s},\mathbf{a})\,\mathrm{d}\mathbf{s}'\,\mathrm{d}\mathbf{z}'\right)\beta(\mathbf{z},\mathbf{b})\,\mathrm{d}\mathbf{z}\,\mathrm{d}\mathbf{b}}{\int_{\mathcal{S},\mathcal{A}} \psi(\mathbf{s},\mathbf{z})\varphi(\mathbf{a},\mathbf{b})\beta(\mathbf{z},\mathbf{b})\,\mathrm{d}\mathbf{z}\,\mathrm{d}\mathbf{b}} \right| \\[2mm]
&\leq \gamma_c \max_{\mathbf{a},\mathbf{s}'} \left| \frac{\int_{\mathcal{S}\times\mathcal{A}} \psi(\mathbf{s},\mathbf{z})\varphi(\mathbf{a},\mathbf{b})\left(\int_{\mathcal{S}} \overline{V}(\mathbf{z}')\phi(\mathbf{z}',\mathbf{s}') - V^*(\mathbf{s}')\,\mathrm{d}\mathbf{z}'\right)\beta(\mathbf{z},\mathbf{b})\,\mathrm{d}\mathbf{z}\,\mathrm{d}\mathbf{b}}{\int_{\mathcal{S},\mathcal{A}} \psi(\mathbf{s},\mathbf{z})\varphi(\mathbf{a},\mathbf{b})\beta(\mathbf{z},\mathbf{b})\,\mathrm{d}\mathbf{z}\,\mathrm{d}\mathbf{b}} \right| \\[2mm]
&= \gamma_c \max_{\mathbf{a},\mathbf{s}'} \left| \frac{\int_{\mathcal{S}\times\mathcal{A}} \psi(\mathbf{s},\mathbf{z})\varphi(\mathbf{a},\mathbf{b})\beta(\mathbf{z},\mathbf{b})\,\mathrm{d}\mathbf{z}\,\mathrm{d}\mathbf{b}}{\int_{\mathcal{S},\mathcal{A}} \psi(\mathbf{s},\mathbf{z})\varphi(\mathbf{a},\mathbf{b})\beta(\mathbf{z},\mathbf{b})\,\mathrm{d}\mathbf{z}\,\mathrm{d}\mathbf{b}} \left(\int_{\mathcal{S}} \overline{V}(\mathbf{z}')\phi(\mathbf{z}',\mathbf{s}') - V^*(\mathbf{s}')\,\mathrm{d}\mathbf{z}'\right) \right| \\[2mm]
&= \gamma_c \max_{\mathbf{s}'} \left| \int_{\mathcal{S}} \overline{V}(\mathbf{z}')\phi(\mathbf{z}',\mathbf{s}') - V^*(\mathbf{s}')\,\mathrm{d}\mathbf{z}' \right| \\[2mm]
&= \gamma_c \max_{\mathbf{s}'} \left| \int_{\mathcal{S}} \overline{V}(\mathbf{s}'+\boldsymbol{\delta})\phi(\mathbf{s}'+\boldsymbol{\delta},\mathbf{s}') - V^*(\mathbf{s}')\,\mathrm{d}\boldsymbol{\delta} \right|.
\end{aligned}
\tag{B.13}
$$

Note that

$$
\phi(\mathbf{s}'+\boldsymbol{\delta},\mathbf{s}') = \prod_{i=1}^{d_s} \frac{e^{-\frac{\delta_i^2}{2h_{\phi,i}^2}}}{\sqrt{2\pi h_{\phi,i}^2}},
$$

thus, using the Lipschitz inequality,

$$
\begin{aligned}
\max_{\mathbf{s}'} &\left| \int_{\mathcal{S}} \overline{V}(\mathbf{s}'+\boldsymbol{\delta})\phi(\mathbf{s}'+\boldsymbol{\delta},\mathbf{s}') - V^*(\mathbf{s}')\,\mathrm{d}\boldsymbol{\delta} \right| \\[2mm]
&\leq \max_{\mathbf{s}'}\left|\overline{V}(\mathbf{s}') - V^*(\mathbf{s}')\right| + \int_{\mathcal{S}} L_V\left(\sum_{i=1}^{d_s}|\delta_i|\right)\prod_{i=1}^{d_s}\frac{e^{-\frac{\delta_i^2}{2h_{\phi,i}^2}}}{\sqrt{2\pi h_{\phi,i}^2}}\,\mathrm{d}\boldsymbol{\delta} \\[2mm]
&= \max_{\mathbf{s}'}\left|\overline{V}(\mathbf{s}') - V^*(\mathbf{s}')\right| + L_V\int_{\mathcal{S}}\left(\sum_{i=1}^{d_s}|\delta_i|\right)\prod_{i=1}^{d_s}\frac{e^{-\frac{\delta_i^2}{2h_{\phi,i}^2}}}{\sqrt{2\pi h_{\phi,i}^2}}\,\mathrm{d}\boldsymbol{\delta} \\[2mm]
&= \max_{\mathbf{s}'}\left|\overline{V}(\mathbf{s}') - V^*(\mathbf{s}')\right| + L_V\sum_{k=1}^{d_s}\left(\prod_{i\neq k}^{d_s}\int_{-\infty}^{+\infty}\frac{e^{-\frac{\delta_i^2}{2h_{\phi,i}^2}}}{\sqrt{2\pi h_{\phi,i}^2}}\,\mathrm{d}\delta_i\right)\int_{-\infty}^{+\infty}|\delta_k|\frac{e^{-\frac{\delta_k^2}{2h_{\phi,k}^2}}}{\sqrt{2\pi h_{\phi,k}^2}}\,\mathrm{d}\delta_k \\[2mm]
&= \max_{\mathbf{s}'}\left|\overline{V}(\mathbf{s}') - V^*(\mathbf{s}')\right| + L_V 2\sum_{k=1}^{d_s}\int_0^{+\infty}\delta_k\frac{e^{-\frac{\delta_k^2}{2h_{\phi,k}^2}}}{\sqrt{2\pi h_{\phi,k}^2}}\,\mathrm{d}\delta_k \\[2mm]
&= \max_{\mathbf{s}'}\left|\overline{V}(\mathbf{s}') - V^*(\mathbf{s}')\right| + L_V\sum_{k=1}^{d_s}\frac{h_{\phi,k}}{\sqrt{2\pi}},
\end{aligned}
$$

which means that

$$
\left|\overline{V}(\mathbf{s}) - V^*(\mathbf{s})\right| \leq A_{\text{Bias}} + \gamma_c\left(\max_{\mathbf{s}'}\left|\overline{V}(\mathbf{s}') - V^*(\mathbf{s}')\right| + L_V\sum_{k=1}^{d_s}\frac{h_{\phi,k}}{\sqrt{2\pi}}\right).
$$

Since both $\overline{V}(\mathbf{s})$ and $V^*(\mathbf{s})$ are bounded by $-R_{\max}/(1-\gamma_c)$ and $R_{\max}(1-\gamma_c)$, then $\left|\overline{V}(\mathbf{s}) - V^*(\mathbf{s})\right| \le 2\frac{R_{\max}}{1-\gamma_c}$, thus

$$\left|\overline{V}(\mathbf{s}) - V^*(\mathbf{s})\right| \le A_{\text{Bias}} + \gamma_c \left(\max_{\mathbf{s}'} \left|\overline{V}(\mathbf{s}') - V^*(\mathbf{s}')\right| + L_V \sum_{k=1}^{d_s} \frac{h_{\phi,k}}{\sqrt{2\pi}} \right) \tag{B.14}$$

$$\left|\overline{V}(\mathbf{s}) - V^*(\mathbf{s})\right| \le A_{\text{Bias}} + \gamma_c \left(2\frac{R_{\max}}{1-\gamma_c} + L_V \sum_{k=1}^{d_s} \frac{h_{\phi,k}}{\sqrt{2\pi}} \right)$$

$$\implies \left|\overline{V}(\mathbf{s}) - V^*(\mathbf{s})\right| \le A_{\text{Bias}} + \gamma_c \left(A_{\text{Bias}} + \gamma_c \left(2\frac{R_{\max}}{1-\gamma_c} + L_V \sum_{k=1}^{d_s} \frac{h_{\phi,k}}{\sqrt{2\pi}} \right) + L_V \sum_{k=1}^{d_s} \frac{h_{\phi,k}}{\sqrt{2\pi}} \right) \qquad \text{using Eq. B.14}$$

$$\implies \left|\overline{V}(\mathbf{s}) - V^*(\mathbf{s})\right| \le \sum_{t=0}^{\infty} \gamma_c^t \left(A_{\text{Bias}} + \gamma_c L_V \sum_{k=1}^{d_s} \frac{h_{\phi,k}}{\sqrt{2\pi}} \right) \qquad \text{using Eq. B.14}$$

$$\implies \left|\overline{V}(\mathbf{s}) - V^*(\mathbf{s})\right| \le \frac{1}{1-\gamma_c} \left(A_{\text{Bias}} + \gamma_c L_V \sum_{k=1}^{d_s} \frac{h_{\phi,k}}{\sqrt{2\pi}} \right).$$

$\square$

B.3. Support to the Empirical Analysis

In this appendix we detail configurations and hyper-parameters used in the empirical analysis of Chapter 5.

B.3.1. Gradient Analysis

The parameters used for the LQG are

$$A = \begin{bmatrix} 1.2 & 0 \\ 0 & 1.1 \end{bmatrix}; \quad B = \begin{bmatrix} 1 & 0 \\ 0 & 1 \end{bmatrix}; \quad Q = \begin{bmatrix} 1 & 0 \\ 0 & 1 \end{bmatrix}; \quad R = \begin{bmatrix} 0.1 & 0 \\ 0 & 0.1 \end{bmatrix}; \quad \Sigma = \begin{bmatrix} 1 & 0 \\ 0 & 1 \end{bmatrix}; \quad \mathbf{s}_0 = [-1, -1].$$

The discount factor is $\gamma = 0.9$, and the length of the episodes is 50 steps. The parameters of the optimization policy are $\theta = [-0.6, -0.8]$ and the off-policy parameters are $\theta' = [-0.35, -0.5]$. The confidence intervals have been computed using *bootstrapped percentile intervals*. The size of the bootstrapped dataset vary from plot to plot (usually from 1000 to 5000 different seeds). The confidence intervals, instead, have been computed using 10000 bootstraps. We used this method instead the more classic standard error (using a χ^2 or a t-distribution), because often, due to the importance sampling, our samples are highly non-Gaussian and heavy-tailed. The bootstrapping method relies on less assumptions, and their confidence intervals were more precise in this case.

B.3.2. Policy Improvement analysis

We use a policy encoded as neural network with parameters θ. A deterministic policy is encoded with a neural network $\mathbf{a} = f_\theta(\mathbf{s})$. The stochastic policy is encoded as a Gaussian distribution with parameters determined by a neural network with two outputs, the mean and covariance. In this case we represent by $f_\theta(\mathbf{s})$ the slice of the output corresponding to the mean and by $g_\theta(\mathbf{s})$ the part of the output corresponding to the covariance.

NOPG can be described with the following hyper-parameters

NOPG Parameters	Meaning
dataset sizes	number of samples contained in the dataset used for training
discount factor γ	usual discount factor in infinite horizon MDP
state $\vec{h}_{\text{factor}}$	constant used to decide the bandwidths for the state-space
action $\vec{h}_{\text{factor}}$	constant used to decide the bandwidths for the action-space
policy	parametrization of the policy
policy output	how is the output of the policy encoded
learning rate	the learning rate and the gradient ascent algorithm used
N_π^{MC} (NOPG-S)	number of samples drawn to compute the integral $\varepsilon_\pi(\mathbf{s})$ with MonteCarlo sampling
N_ϕ^{MC}	number of samples drawn to compute the integral over the next state $\int \phi(\mathbf{s}')\,\mathrm{d}\mathbf{s}'$
$N_{\mu_0}^{\text{MC}}$	number of samples drawn to compute the integral over the initial distribution $\int \hat{V}_\pi(\mathbf{s})\mu_0(\mathbf{s})\,\mathrm{d}\mathbf{s}$
policy updates	number of policy updates before returning the optimized policy

A few considerations about NOPG parameters. If $N_\phi^{\text{MC}} = 1$ we use the mean of the kernel ϕ as a sample to approximate the integral over the next state. When optimizing a stochastic policy represented by a Gaussian distribution, we set and linearly decay the variance over the policy optimization procedure. The kernel bandwidths are computed in two steps: first we find the best bandwidth for each dimension of the state and action spaces using cross validation; second we multiply each bandwidth by an empirical constant factor ($\vec{h}_{\text{factor}}$). This second step is important to guarantee that the state and action spaces do not have a zero density. For instance, in a continuous action environment, when sampling actions from a uniform grid we have to guarantee that the space between the grid points have some density. The problem of estimating the bandwidth in kernel density estimation is well studied, but needs to be adapted to the problem at hand, specially with a low number of samples. We found this approach to work well for our experiments but it can be further improved.

Pendulum with Uniform Dataset. Tables B.2 and B.3 describe the hyper-parameters used to run the experiment shown in the first plot of Figure 5.7.

Dataset Generation: The datasets have been generated using a grid over the state-action spaces $\theta, \dot{\theta}, u$, where θ and $\dot{\theta}$ are respectively angle and angular velocity of the pendulum, and u is the torque applied. In Table B.2 are enumerated the different datasets used.

$\#\theta$	$\#\dot\theta$	$\#u$	Sample size
10	10	2	200
15	15	2	450
20	20	2	800
25	25	2	1250
30	30	2	1800
40	40	2	3200

Table B.2.: **Pendulum uniform grid dataset configurations** This table shows the level of discretization for each dimension of the state space ($\#\theta$ and $\#\dot\theta$) and the action space ($\#u$). Each line corresponds to a uniformly sampled dataset, where $\theta \in [-\pi, \pi]$, $\dot\theta \in [-8, 8]$ and $u \in [-2, 2]$. The entries under the states' dimensions and action dimension correspond to how many linearly spaced states or actions are to be queried from the corresponding intervals. The Cartesian product of states and actions dimensions is taken in order to generate the state-action pairs to query the environment transitions. The rightmost column indicates the total number of corresponding samples.

Algorithm details: The configuration used for NOPG-D and NOPG-S are listed in Table B.3.

NOPG	
discount factor γ	0.97
state $\vec{h}_{\text{factor}}$	1.0 1.0 1.0
action $\vec{h}_{\text{factor}}$	50.0
policy	neural network parameterized by $\boldsymbol{\theta}$
	1 hidden layer, 50 units, ReLU activations
policy output	$2\tanh(f_{\boldsymbol{\theta}}(\mathbf{s}))$ (NOGP-D)
	$\mu = 2\tanh(f_{\boldsymbol{\theta}}(\mathbf{s}))$, $\sigma = \text{sigmoid}(g_{\boldsymbol{\theta}}(\mathbf{s}))$ (NOGP-S)
learning rate	10^{-2} with ADAM optimizer
N_{π}^{MC} (NOPG-S)	15
N_{ϕ}^{MC}	1
$N_{\mu_0}^{\text{MC}}$	(non applicable) fixed initial state
policy updates	$1.5 \cdot 10^3$

Table B.3.: **NOPG configurations for the Pendulum uniform grid experiment**

Pendulum with Random Agent. The following tables show the hyper-parameters used for generating the second plot starting from the left in Figure 5.7

NOPG	
dataset sizes	$10^2, 5 \cdot 10^2, 10^3, 1.5 \cdot 10^3, 2 \cdot 10^3, 3 \cdot 10^3,$
	$5 \cdot 10^3, 7 \cdot 10^3, 9 \cdot 10^3, 10^4$
discount factor γ	0.97
state $\vec{h}_{\text{factor}}$	1.0 1.0 1.0
action $\vec{h}_{\text{factor}}$	25.0
policy	neural network parameterized by $\boldsymbol{\theta}$

	1 hidden layer, 50 units, ReLU activations
policy output	$2\tanh(f_{\boldsymbol{\theta}}(\mathbf{s}))$ (NOGP-D)
	$\mu = 2\tanh(f_{\boldsymbol{\theta}}(\mathbf{s}))$, $\sigma = \mathrm{sigmoid}(g_{\boldsymbol{\theta}}(\mathbf{s}))$ (NOGP-S)
learning rate	10^{-2} with ADAM optimizer
$N_{\pi_{\theta}}^{\mathrm{MC}}$ (NOPG-S)	10
N_{ϕ}^{MC}	1
$N_{\mu_0}^{\mathrm{MC}}$	(non applicable) fixed initial state
policy updates	$2 \cdot 10^3$

SAC

discount factor γ	0.97
rollout steps	500
actor	neural network parameterized by θ_{actor}
	1 hidden layer, 50 units, ReLU activations
actor output	$2\tanh(u)$, $u \sim \mathcal{N}(\cdot\,|\mu = f_{\boldsymbol{\theta}_{\mathrm{actor}}}(\mathbf{s}), \sigma = g_{\boldsymbol{\theta}_{\mathrm{actor}}}(\mathbf{s}))$
actor learning rate	10^{-3} with ADAM optimizer
critic	neural network parameterized by θ_{critic}
	2 hidden layers, 50 units, ReLU activations
critic output	$f_{\boldsymbol{\theta}_{\mathrm{critic}}}(\mathbf{s}, \mathbf{a})$
critic learning rate	$5 \cdot 10^{-3}$ with ADAM optimizer
max replay size	$5 \cdot 10^5$
initial replay size	128
batch size	64
soft update	$\tau = 5 \cdot 10^{-3}$
policy updates	$2.5 \cdot 10^5$

DDPG / TD3

discount factor γ	0.97
rollout steps	500
actor	neural network parameterized by θ_{actor}
	1 hidden layer, 50 units, ReLU activations
actor output	$2\tanh(f_{\boldsymbol{\theta}_{\mathrm{actor}}}(\mathbf{s}))$
actor learning rate	10^{-3} with ADAM optimizer
critic	neural network parameterized by θ_{critic}
	2 hidden layers, 50 units, ReLU activations
critic output	$f_{\boldsymbol{\theta}_{\mathrm{critic}}}(\mathbf{s}, \mathbf{a})$
critic learning rate	10^{-2} with ADAM optimizer
max replay size	$5 \cdot 10^5$
initial replay size	128
batch size	64
soft update	$\tau = 5 \cdot 10^{-3}$
policy updates	$2.5 \cdot 10^5$

DDPG Offline

dataset sizes	$10^2, 5 \cdot 10^2, 10^3, 2 \cdot 10^3, 5 \cdot 10^3, 7.5 \cdot 10^3,$ $10^4, 1.2 \cdot 10^4, 1.5 \cdot 10^4, 2 \cdot 10^4, 2.5 \cdot 10^4$
discount factor γ	0.97
actor	neural network parameterized by θ_{actor} 1 hidden layer, 50 units, ReLU activations
actor output	$2 \tanh(f_{\theta_{\text{actor}}}(\mathbf{s}))$
actor learning rate	10^{-2} with ADAM optimizer
critic	neural network parameterized by θ_{critic} 1 hidden layer, 50 units, ReLU activations
critic output	$f_{\theta_{\text{critic}}}(\mathbf{s}, \mathbf{a})$
critic learning rate	10^{-2} with ADAM optimizer
soft update	$\tau = 10^{-3}$
policy updates	$2 \cdot 10^3$

Table B.4.: **Algorithms configurations for the Pendulum random data experiment**

Cart-pole with Random Agent. The following tables show the hyper-parameters used to generate the third plot in Figure 5.7.

NOPG

dataset sizes	$10^2, 2.5 \cdot 10^2, 5 \cdot 10^2, 10^3, 1.5 \cdot 10^3, 2.5 \cdot 10^3,$ $3 \cdot 10^3, 5 \cdot 10^3, 6 \cdot 10^3, 8 \cdot 10^3, 10^4$
discount factor γ	0.99
state $\vec{h}_{\text{factor}}$	1.0 1.0 1.0
action $\vec{h}_{\text{factor}}$	20.0
policy	neural network parameterized by θ 1 hidden layer, 50 units, ReLU activations
policy output	$5 \tanh(f_\theta(\mathbf{s}))$ (NOGP-D) $\mu = 5 \tanh(f_\theta(\mathbf{s})), \sigma = \text{sigmoid}(g_\theta(\mathbf{s}))$ (NOGP-S)
learning rate	10^{-2} with ADAM optimizer
N_π^{MC} (NOPG-S)	10
N_ϕ^{MC}	1
$N_{\mu_0}^{\text{MC}}$	15
policy updates	$2 \cdot 10^3$

SAC

discount factor γ	0.99
rollout steps	10000
actor	neural network parameterized by θ_{actor} 1 hidden layer, 50 units, ReLU activations
actor output	$5 \tanh(u), u \sim \mathcal{N}(\cdot \mid \mu = f_{\theta_{\text{actor}}}(\mathbf{s}), \sigma = g_{\theta_{\text{actor}}}(\mathbf{s}))$
actor learning rate	10^{-3} with ADAM optimizer

critic	neural network parameterized by θ_{critic}
	2 hidden layers, 50 units, ReLU activations
critic output	$f_{\theta_{\text{critic}}}(\mathbf{s}, \mathbf{a})$
critic learning rate	$5 \cdot 10^{-3}$ with ADAM optimizer
max replay size	$5 \cdot 10^5$
initial replay size	128
batch size	64
soft update	$\tau = 5 \cdot 10^{-3}$
policy updates	$2.5 \cdot 10^5$

DDPG / TD3

discount factor γ	0.99
rollout steps	10000
actor	neural network parameterized by θ_{actor}
	1 hidden layer, 50 units, ReLU activations
actor output	$5 \tanh(f_{\theta_{\text{actor}}}(\mathbf{s}))$
actor learning rate	10^{-3} with ADAM optimizer
critic	neural network parameterized by θ_{critic}
	1 hidden layer, 50 units, ReLU activations
critic output	$f_{\theta_{\text{critic}}}(\mathbf{s}, \mathbf{a})$
critic learning rate	10^{-2} with ADAM optimizer
soft update	$\tau = 10^{-3}$
policy updates	$2 \cdot 10^5$

DDPG Offline

dataset sizes	$10^2, 5 \cdot 10^2, 10^3, 2 \cdot 10^3, 3.5 \cdot 10^3, 5 \cdot 10^3,$
	$8 \cdot 10^3, 10^4, 1.5 \cdot 10^4, 2 \cdot 10^4, 2.5 \cdot 10^4$
discount factor γ	0.99
actor	neural network parameterized by θ_{actor}
	1 hidden layer, 50 units, ReLU activations
actor output	$5 \tanh(f_{\theta_{\text{actor}}}(\mathbf{s}))$
actor learning rate	10^{-2} with ADAM optimizer
critic	neural network parameterized by θ_{critic}
	1 hidden layer, 50 units, ReLU activations
critic output	$f_{\theta_{\text{critic}}}(\mathbf{s}, \mathbf{a})$
critic learning rate	10^{-2} with ADAM optimizer
soft update	$\tau = 10^{-3}$
policy updates	$2 \cdot 10^3$

Table B.5.: **Algorithms configurations for the CartPole random data experiment.**

Mountain Car with Human Demonstrator.   The dataset (10 trajectories) for the experiment in Figure 5.9 has been generated by a human demonstrator, and is available in the source code provided.

NOPG	
discount factor γ	0.99
state $\vec{h}_{\text{factor}}$	1.0 1.0
action $\vec{h}_{\text{factor}}$	50.0
policy	neural network parameterized by $\vec{\theta}$
	1 hidden layer, 50 units, ReLU activations
policy output	$1 \tanh(f_{\vec{\theta}}(\mathbf{s}))$ (NOGP-D)
	$\mu = 1 \tanh(f_{\vec{\theta}}(\mathbf{s}))$, $\sigma = \text{sigmoid}(g_{\vec{\theta}}(\mathbf{s}))$ (NOGP-S)
learning rate	10^{-2} with ADAM optimizer
N_{π}^{MC} (NOPG-S)	15
N_{ϕ}^{MC}	1
$N_{\mu 0}^{\text{MC}}$	15
policy updates	$1.5 \cdot 10^{3}$

Table B.6.: **NOPG configurations for the MountainCar experiment.**

C. An Upper Bound on the Bias of Natadaya-Watson Kernel Regression

In order to give the proof of the stated Theorems, we need to introduce some quantities and to state some facts that will be used in our proofs.

C.1. Preliminaries

In order to provide proofs of the stated Theorems, we need to introduce some quantities and to state some facts that will be used in our proofs.

Proposition 5. *With $a \in \mathbb{R}$, $h \neq 0$ and $c \in \mathbb{R}$,*

$$\int_0^a k_i\left(\frac{x}{h}\right) e^{-xL_f}\,\mathrm{d}x = B_i(a/h, ch). \tag{C.1}$$

Proposition 6. *With $a, b \geq 0$, $h_i > 0$ and $L_f \geq 0$,*

$$\int_{-a}^b k_i\left(\frac{x}{h_i}\right) e^{-|x|L_f}\,\mathrm{d}x = h_i B_i(b/h_i, L_f h_i) - h_i B_i(-a/h, -L_f h_i) = \psi_i(a, b). \tag{C.2}$$

Proposition 7. *With $a, b \geq 0$, $h_i > 0$ and $L_f \geq 0$,*

$$\int_{-a}^b k_i\left(\frac{x}{h_i}\right) e^{|x|L_f}\,\mathrm{d}x = h_i B_i(b/h_i, -L_f h_i L_f) - h_i B_i(-a/h_i, L_f h_i) = \zeta_i(a, b). \tag{C.3}$$

Proposition 8. *With $a \in \mathbb{R}$, $h \neq 0$ and $c \in \mathbb{R}$,*

$$\int_0^a k_i\left(\frac{x}{h}\right) e^{-xc} x\,\mathrm{d}x = h_i^2\left(C_i(a/h, ch)\right). \tag{C.4}$$

Proposition 9. *With $a, b \geq 0$, $h_i > 0$ and $L_f \geq 0$,*

$$\int_{-a}^b k\left(\frac{x}{h_i}\right) e^{|x|L_f} |x|\,\mathrm{d}x = h_i^2\left(C_i(b/h, L_f h_i) + C_i(-a/h_i, L_f h_i)\right) = \xi_i(a, b). \tag{C.5}$$

Definition 7. Integral on a d-interval

Let $\mathcal{C} \equiv \Omega(\boldsymbol{\tau}^-, \boldsymbol{\tau}^+)$ with $\boldsymbol{\tau}^-, \boldsymbol{\tau}^+ \in \overline{\mathbb{R}}^d$. Let the integral of a function $f : \mathcal{C} \to \mathbb{R}$ defined on $\mathcal{C}$ be defined as

$$\int_{\mathcal{C}} f(\mathbf{x})\,\mathrm{d}\mathbf{x} = \int_{\tau_1^-}^{\tau_1^+} \int_{\tau_2^-}^{\tau_2^+} \cdots \int_{\tau_d^-}^{\tau_d^+} f([x_1, x_2, \ldots, x_d]^{\mathsf{T}})\,\mathrm{d}x_d \ldots \mathrm{d}x_2\,\mathrm{d}x_1.$$

Proposition 10. *There is a function* $g : \Upsilon \to \mathbb{R}$ *such that*

$$f_X(\mathbf{x}) = \frac{e^{g(\mathbf{x})}}{\int_{\Upsilon} e^{g(\mathbf{x})}\,\mathrm{d}\mathbf{x}}$$

and, given A2, $|g(\mathbf{x}) - g(\mathbf{y})| \le L_f |\mathbf{x} - \mathbf{y}| \quad \forall \mathbf{y} \in \mathcal{D}.$

Proposition 11. Independent Factorization

Let $\mathcal{C} \equiv \Omega(\boldsymbol{\tau}^-, \boldsymbol{\tau}^+)$ *where* $\boldsymbol{\tau}^-, \boldsymbol{\tau}^+ \in \mathbb{R}^d$, *and* $f_i : \mathbb{R} \to \mathbb{R}$,

$$\int_{\mathcal{C}} \prod_{i=1}^{d} f_i(x_i)\,\mathrm{d}\mathbf{x} = \prod_{i=1}^{d} \int_{\mathcal{C}} f_i(x_i)\,\mathrm{d}\mathbf{x}.$$

Proposition 12. *Given* $\mathcal{C} \equiv \Omega(\boldsymbol{\tau}^-, \boldsymbol{\tau}^+)$, $p : \mathbb{R} \to \mathbb{R}$, $q : \mathbb{R} \to \mathbb{R}$,

$$\int_{\mathcal{C}} \left(\prod_{i=1}^{d} p(z_i) \right) \left(\sum_{k=1}^{d} g(z_k) \right) \mathrm{d}\mathbf{z} = \sum_{k=1}^{d} \left(\prod_{i \neq k}^{d} \int_{\tau_i^-}^{\tau_i^+} p(z)\,\mathrm{d}z \right) \int_{\tau_k^-}^{\tau_k^+} p(z)q(z)\,\mathrm{d}z.$$

C.2. Proof of Theorem 5

Proof. Proof of Theorem 5:

$$\left| \mathbb{E}\left[\lim_{n \to \infty} \hat{f}_n(\mathbf{x}) \right] - m(\mathbf{x}) \right|$$

$$= \left| \mathbb{E}\left[\lim_{n \to \infty} \frac{\sum_{i=1}^{n} K_{\mathbf{h}}(\mathbf{x} - \mathbf{x}_i) y_i}{\sum_{j=1}^{n} K_{\mathbf{h}}(\mathbf{x} - \mathbf{x}_j)} \right] - m(\mathbf{x}) \right|$$

$$= \left| \mathbb{E}\left[\lim_{n \to \infty} \frac{n^{-1} \sum_{i=1}^{n} K_{\mathbf{h}}(\mathbf{x} - \mathbf{x}_i) y_i}{n^{-1} \sum_{j=1}^{n} K_{\mathbf{h}}(\mathbf{x} - \mathbf{x}_j)} \right] - m(\mathbf{x}) \right|$$

$$= \left| \mathbb{E}\left[\frac{\int_{\Upsilon} K_{\mathbf{h}}(\mathbf{x} - \mathbf{z})(m(\mathbf{z}) - \epsilon(\mathbf{z})) f_X(\mathbf{z})\,\mathrm{d}\mathbf{z}}{\int_{\Upsilon} K_{\mathbf{h}}(\mathbf{x} - \mathbf{z}) f_X(\mathbf{z})\,\mathrm{d}\mathbf{z}} \right] - m(\mathbf{x}) \right|$$

$$= \left| \frac{\int_{\Upsilon} K_{\mathbf{h}}(\mathbf{x} - \mathbf{z}) m(\mathbf{z}) f_X(\mathbf{z})\,\mathrm{d}\mathbf{z}}{\int_{\Upsilon} K_{\mathbf{h}}(\mathbf{x} - \mathbf{z}) f_X(\mathbf{z})\,\mathrm{d}\mathbf{z}} - m(\mathbf{x}) \right|$$

$$= \left| \frac{\int_{\Upsilon} K_{\mathbf{h}}(\mathbf{x} - \mathbf{z})(m(\mathbf{z}) - m(\mathbf{x})) f_X(\mathbf{z})\,\mathrm{d}\mathbf{z}}{\int_{\Upsilon} K_{\mathbf{h}}(\mathbf{x} - \mathbf{z}) f_X(\mathbf{z})\,\mathrm{d}\mathbf{z}} \right|$$

$$= \frac{\left| \int_{\Upsilon} K_{\mathbf{h}}(\mathbf{x} - \mathbf{z})(m(\mathbf{z}) - m(\mathbf{x})) f_X(\mathbf{z})\,\mathrm{d}\mathbf{z} \right|}{\left| \int_{\Upsilon} K_{\mathbf{h}}(\mathbf{x} - \mathbf{z}) f_X(\mathbf{z})\,\mathrm{d}\mathbf{z} \right|}.$$

We want to obtain an upper bound of the bias. Therefore we want to find an upper bound of the numerator and a lower bound of the denominator.

Lower bound of the Denominator:
The denominator is always positive, so the module can be removed,

$$\int_{\Upsilon} K_{\mathbf{h}}(\mathbf{x} - \mathbf{z}) f_X(\mathbf{z}) \, d\mathbf{z}$$

$$= \int_{\Upsilon} f_X(\mathbf{z}) \prod_{i=1}^{d} k\left(\frac{z_i}{h_i}\right) d\mathbf{z}$$

$$\geq \int_{\mathcal{D}} f_X(\mathbf{z}) \prod_{i=1}^{d} k\left(\frac{z_i}{h_i}\right) d\mathbf{z} \quad \text{(since } \mathcal{D} \subseteq \Upsilon \text{ and the integrand is always non-negative)}$$

$$= \frac{e^{g(\mathbf{x})}}{\int_{\Upsilon} e^{g(\mathbf{z})} \, d\mathbf{z}} \int_{\mathcal{D}} e^{g(\mathbf{z}) - g(\mathbf{x})} \prod_{i=1}^{d} k\left(\frac{z_i}{h_i}\right) d\mathbf{z} \qquad \text{(Prop. 10)}$$

$$= f_X(\mathbf{x}) \int_{\overline{\mathcal{D}}} e^{g(\mathbf{x}+\mathbf{l}) - g(\mathbf{x})} \prod_{i=1}^{d} k\left(\frac{l_i}{h_i}\right) d\mathbf{l} \qquad \text{let } \mathbf{l} = \mathbf{z} - \mathbf{x} \text{ and } \overline{\mathcal{D}} \equiv \Omega(-\delta^-, +\delta^+)$$

$$\geq f_X(\mathbf{x}) \int_{\overline{\mathcal{D}}} e^{-|\mathbf{l}| L_f} \prod_{i=1}^{d} k\left(\frac{l_i}{h_i}\right) d\mathbf{l} \qquad \text{(Axiom A2 + Lipschitz Inequality)}$$

$$= f_X(\mathbf{x}) \int_{\overline{\mathcal{D}}} \prod_{i=1}^{d} e^{-|l_i| L_f} k\left(\frac{l_i}{h_i}\right) d\mathbf{l}$$

Now considering Propositions 5 and 11, we obtain

$$\int_{-\infty}^{+\infty} K_{\mathbf{h}}(\mathbf{x} - \mathbf{z}) f_X(\mathbf{z}) \, d\mathbf{z} \geq f_X(\mathbf{x}) \prod_{i=1}^{d} \psi_i(\delta_i^-, \delta_i^+). \qquad (C.6)$$

Upper bound of the Numerator:

$$\left| \int_{\Upsilon} K_{\mathbf{h}}(\mathbf{x} - \mathbf{z}) (m(\mathbf{z}) - m(\mathbf{x})) f_X(\mathbf{z}) \, d\mathbf{z} \right|$$

$$\leq \int_{\Upsilon} K_{\mathbf{h}}(\mathbf{x} - \mathbf{z}) |m(\mathbf{z}) - m(\mathbf{x})| f_X(\mathbf{z}) \, d\mathbf{z}$$

$$= \int_{\mathcal{G}} K_{\mathbf{h}}(\mathbf{x} - \mathbf{z}) |m(\mathbf{z}) - m(\mathbf{x})| f_X(\mathbf{z}) \, d\mathbf{z} + \int_{\Upsilon \setminus \mathcal{G}} K_{\mathbf{h}}(\mathbf{x} - \mathbf{z}) |m(\mathbf{z}) - m(\mathbf{x})| f_X(\mathbf{z}) \, d\mathbf{z}$$

$$\leq \int_{\mathcal{G}} K_{\mathbf{h}}(\mathbf{x} - \mathbf{z}) |m(\mathbf{z}) - m(\mathbf{x})| f_X(\mathbf{z}) \, d\mathbf{z} + f_X(\mathbf{x}) M \int_{\Upsilon \setminus \mathcal{G}} K_{\mathbf{h}}(\mathbf{x} - \mathbf{z}) \, d\mathbf{z}$$

$$= \frac{e^{g(\mathbf{x})}}{\int_{\Upsilon} e^{g(\mathbf{z})} \, d\mathbf{z}} \int_{\mathcal{G}} e^{g(\mathbf{z}) - g(\mathbf{x})} K_{\mathbf{h}}(\mathbf{x} - \mathbf{z}) |m(\mathbf{z}) - m(\mathbf{x})| \, d\mathbf{z} + f_X(\mathbf{x}) M \int_{\Upsilon \setminus \mathcal{G}} K_{\mathbf{h}}(\mathbf{x} - \mathbf{z}) \, d\mathbf{z}$$

$$\leq f_X(\mathbf{x}) \left(\int_{\mathcal{G}} e^{g(\mathbf{z}) - g(\mathbf{x})} K_{\mathbf{h}}(\mathbf{x} - \mathbf{z}) |m(\mathbf{z}) - m(\mathbf{x})| \, d\mathbf{z} + MD \right)$$

where

$$D = \int_{\Upsilon \setminus \mathcal{G}} K_{\mathbf{h}}(\mathbf{x} - \mathbf{z})\, \mathrm{d} = \lim_{\omega \to +\infty} \prod_{i=1}^{d} \int_{x_i - \omega}^{x_i + \omega} k_i \left(\frac{x_i - z}{h_i} \right) \mathrm{d}z - \prod_{i=1}^{d} \int_{x_i - \gamma_i^-}^{x_i + \gamma_i^+} k_i \left(\frac{x_i - z}{h_i} \right) \mathrm{d}z$$

$$= \lim_{\omega \to +\infty} \prod_{i=1}^{d} h_i \Phi_i \left(\frac{\omega}{h_i} \right) - \prod_{i=1}^{d} h_i \Phi_i \left(\frac{\gamma_i^+}{h} \right) + h_i \Phi_i \left(\frac{\gamma_i^-}{h_i} \right). \tag{C.7}$$

Let $\mathcal{F} \equiv \Omega(\mathbf{x} - \boldsymbol{\phi}^-, \mathbf{x} + \boldsymbol{\phi}^+) \subseteq \mathcal{G}$, we will later define at our convenience.

$$f_X(\mathbf{x}) \left(\int_{\mathcal{G}} e^{g(\mathbf{z}) - g(\mathbf{x})} K_{\mathbf{h}}(\mathbf{x} - \mathbf{z})\, |m(\mathbf{z}) - m(\mathbf{x})|\, \mathrm{d}\mathbf{z} + MD \right)$$

$$= f_X(\mathbf{x}) \left(\int_{\mathcal{F}} e^{g(\mathbf{z}) - g(\mathbf{x})} K_{\mathbf{h}}(\mathbf{x} - \mathbf{z})\, |m(\mathbf{z}) - m(\mathbf{x})|\, \mathrm{d}\mathbf{z} + \right.$$

$$\left. \int_{\mathcal{G} \setminus \mathcal{F}} e^{g(\mathbf{z}) - g(\mathbf{x})} K_{\mathbf{h}}(\mathbf{x} - \mathbf{z})\, |m(\mathbf{z}) - m(\mathbf{x})|\, \mathrm{d}\mathbf{z} + MD \right)$$

$$\leq f_X(\mathbf{x}) \left(\int_{\mathcal{F}} e^{g(\mathbf{z}) - g(\mathbf{x})} K_{\mathbf{h}}(\mathbf{x} - \mathbf{z})\, |m(\mathbf{z}) - m(\mathbf{x})|\, \mathrm{d}\mathbf{z} \right.$$

$$\left. + M \int_{\mathcal{G} \setminus \mathcal{F}} e^{g(\mathbf{z}) - g(\mathbf{x})} K_{\mathbf{h}}(\mathbf{x} - \mathbf{z})\, \mathrm{d}\mathbf{z} + MD \right)$$

$$= f_X(\mathbf{x}) \left(\int_{\overline{\mathcal{F}}} e^{g(\mathbf{x} + \mathbf{l}) - g(\mathbf{x})} K_{\mathbf{h}}(-\mathbf{l})\, |m(\mathbf{x} + \mathbf{l}) - m(\mathbf{l})|\, \mathrm{d}\mathbf{l} \right.$$

$$\left. + M \int_{\overline{\mathcal{G}} \setminus \overline{\mathcal{F}}} e^{g(\mathbf{x} + \mathbf{l}) - g(\mathbf{x})} K_{\mathbf{h}}(-\mathbf{l})\, \mathrm{d}\mathbf{l} + MD \right)$$

$$\text{with } \mathbf{l} = \mathbf{z} - \mathbf{x},\ \overline{\mathcal{F}} \equiv \Omega(-\boldsymbol{\phi}^-, \boldsymbol{\phi}^+) \text{ and } \overline{G} \equiv \Omega(-\boldsymbol{\gamma}^-, \boldsymbol{\gamma}^+)$$

$$\leq f_X(\mathbf{x}) \left(\int_{\overline{\mathcal{F}}} e^{L_f |\mathbf{l}|} K_{\mathbf{h}}(\mathbf{l}) L_m\, |\mathbf{l}|\, \mathrm{d}\mathbf{l} + M \int_{\overline{\mathcal{G}} \setminus \overline{\mathcal{F}}} e^{L_f |\mathbf{l}|} K_{\mathbf{h}}(\mathbf{l})\, \mathrm{d}\mathbf{l} + MD \right)$$

$$\text{(A2, A3 + Lipschitz Inequality)}$$

The first integral instead can be solved with Propositions 7, 9 and 12,

$$\int_{\overline{\mathcal{F}}} e^{L_f |\mathbf{l}|} K_{\mathbf{h}}(\mathbf{l}) L_m\, |\mathbf{l}|\, \mathrm{d}\mathbf{l}$$

$$= \int_{\overline{\mathcal{F}}} \left(\prod_{i=1}^{d} k \left(\frac{l}{h_i} \right) e^{|l| L_f} \right) L_m \sum_{i=1}^{d} |l_i|\, \mathrm{d}l\, \mathrm{d}\mathbf{l}$$

$$= L_m \sum_{k=1}^{d} \left(\prod_{i \neq k}^{d} \int_{-\phi_i^-}^{\phi_i^+} k \left(\frac{l}{h_i} \right) e^{|l| L_f}\, \mathrm{d}z \right) \int_{-\phi_k^-}^{\phi_k^+} k \left(\frac{l}{h} \right) e^{|l| L_f} |l_i|\, \mathrm{d}l \tag{Prop. 12}$$

$$= L_m \sum_{k=1}^{d} \xi_k(\phi_k^-, \phi_k^+) \left(\prod_{i \neq k}^{d} \zeta_i(\phi_i^-, \phi_i^+) \right).$$

The second integral can be solved using Propositions 5, 11,

$$\int_{\mathcal{G} \setminus \mathcal{F}} e^{L_f |\mathbf{l}|} K_{\mathbf{h}}(\mathbf{l})\, \mathrm{d}\mathbf{z} = \int_{\mathcal{G}} e^{L_f |\mathbf{l}|} K_{\mathbf{h}}(\mathbf{l})\, \mathrm{d}\mathbf{l} - \int_{\mathcal{F}} e^{L_f |\mathbf{l}|} K_{\mathbf{h}}(\mathbf{l})\, \mathrm{d}\mathbf{l}$$

$$= \prod_{i=1}^{d} \zeta_i(\gamma_i^-, \gamma_i^+) - \prod_{i=1}^{d} \zeta_i(\phi_i^-, \phi_i^+).$$

A good choice for $\mathcal{F}$ is $\phi_i^- = \min(\gamma_i^-, M/L_f)$ and $\phi_i^+ = \min(\gamma_i^+, M/L_f)$, as in this way we obtain a tighter bound. In last analysis,

$$\left| \mathbb{E}\left[\lim_{n \to \infty} \hat{m}_n(\mathbf{x}) \right] - m(\mathbf{x}) \right| \leq \frac{L_m \sum_{k=1}^{d} \xi_k(\phi_i^k, \phi_i^k) \prod_{i \neq k}^{d} \zeta(\phi_i^-, \phi_i^+) + M\left(\prod_{i=1}^{d} \zeta(\gamma_i^-, \gamma_i^+) - \prod_{i=1}^{d} \zeta(\phi_i^-, \phi_i^+) + D \right)}{\prod_{i=1}^{d} \psi_i(\delta_i^-, \delta_i^+)}$$

showing the correctness of Theorem 5. $\qquad\square$

In order to prove Theorem 6 we shall note that $\Upsilon \equiv \mathcal{G} \equiv \mathcal{F}$, therefore the lower bound can be bounded by

$$
\begin{aligned}
\int_{\Upsilon} K_{\mathbf{h}}(\mathbf{x} - \mathbf{z}) f_X(\mathbf{z}) \, \mathrm{d}\mathbf{z} &= \int_{\Upsilon} f_X(\mathbf{z}) \prod_{i=1}^{d} k_i \left(\frac{x_i - z_i}{h_i} \right) \mathrm{d}\mathbf{z} \\
&\geq f_X(\mathbf{x}) \int_{\overline{\Upsilon}} \prod_{i=1}^{d} k_i \left(\frac{l_i}{h_i} \right) e^{-l_i L_f} \, \mathrm{d}\mathbf{l} \\
&= f_X(\mathbf{x}) \prod_{i=1}^{d} \psi_i(v_i^-, v_i^+)
\end{aligned}
\tag{C.8}
$$

for the numerator, instead

$$
\begin{aligned}
&\left| \int_{\Upsilon} K_{\mathbf{h}}(\mathbf{x} - \mathbf{z})\left(m(\mathbf{z}) - m(\mathbf{x}) \right) f_X(\mathbf{z}) \, \mathrm{d}\mathbf{z} \right| \\
&\leq f_X(\mathbf{x}) \int_{\Upsilon} e^{g(\mathbf{z}) - g(\mathbf{x})} K_{\mathbf{h}}(\mathbf{x} - \mathbf{z}) \left| m(\mathbf{z}) - m(\mathbf{x}) \right| \mathrm{d}\mathbf{z} \\
&\leq f_X(\mathbf{x}) \int_{\overline{\Upsilon}} e^{L_f |\mathbf{l}|} K_{\mathbf{h}}(\mathbf{l}) L_m |\mathbf{l}| \, \mathrm{d}\mathbf{l} \qquad \text{where } \overline{\Upsilon} \equiv \Omega(v^-, v^+)
\end{aligned}
$$

and therefore, for the reasoning already made for Theorem 5,

$$\left| \mathbb{E}\left[\lim_{n \to \infty} \hat{m}_n(\mathbf{x}) \right] - m(\mathbf{x}) \right| \leq \frac{L_m \sum_{k=1}^{d} \xi_k(v_k^-, v_k^+) \prod_{i \neq k}^{d} \zeta_i(v_i^-, v_i^+)}{\prod_{i=1}^{d} \psi_i(v_i^-, v_i^+)}$$

where $\xi_{A_k}, \zeta, \Psi, \varphi$ are defined as in Theorem 5 and $\phi_i^- = v_i^-, \phi_i^+ = v_i^+$.

C.3. Kernel Details

In this section we characterize the functions $\phi(x)$, $B(a, c)$ and $C(a, c)$ for Gaussian, box and triangular kernels.

C.3.1. Gaussian Kernel

The Gaussian kernel, defined as $k(x) = e^{-x^2}$, is characterized by,

$$\phi(x) = \frac{\sqrt{\pi}}{2}\operatorname{erf}(a), \quad B(a,c) = \frac{\sqrt{\pi}}{2}e^{\frac{c^2}{4}}\left(\operatorname{erf}\left(a+\frac{c}{2}\right) - \operatorname{erf}\left(\frac{c}{2}\right)\right),$$
$$C(a,c) = \frac{1}{2}\left(1 - e^{-a(a+c)} - cB(a,c)\right).$$

Figure C.1 depicts the results of a numerical simulation using the Gaussian kernel.

C.3.2. Box Kernel

The box kernel, defined as $k(x) = I(x)$, has $B(a,c) = g_B(a,c) - g_B(0,c)$ and $C(a,c) = g_C(a,c) - g_C(0,c)$, where

$$g_B(x,c) = \int k(x)e^{-xc}\,\mathrm{d}x = \begin{cases} 0 & \text{if } x < -1 \\ \frac{e^{-c}\left(e^{2c}-1\right)}{c} & \text{if } x > 1, \ (+\text{const}), \\ \frac{e^{-c}\left(e^{cx+c}-1\right)}{c} & \text{otherwise} \end{cases}$$

$$g_C(x,c) = \int k(x)e^{-xc}x\,\mathrm{d}x = \begin{cases} 0 & \text{if } x < -1 \\ \frac{e^{-c}\left(ce^{2c}-e^{2c}+c+1\right)}{c^2} & \text{if } x > 1, \ (+\text{const}), \\ \frac{e^{-c}\left(cxe^{cx+c}-e^{cx+c}+c+1\right)}{c^2} & \text{otherwise} \end{cases}$$

Figure C.2 depicts the results of a numerical simulation using the Box kernel.

C.3.3. Triangular Kernel

The box kernel, defined as $k(x) = I(x)(1 - |x|)$, has $B(a,c) = g_B(a,c) - g_B(0,c)$ and $C(a,c) = g_C(a,c) - g_C(0,c)$, where

$$g_B(x,c) = \int k(x)e^{-xc}\,\mathrm{d}x = \begin{cases} 0 & \text{if } x < -1 \\ \frac{e^{-cx}\left(e^{cx+c}-c(x+1)-1\right)}{c^2} & \text{if } -1 \leq x \leq 0, \\ \frac{e^{-cx}\left(c(x-1)-2e^{cx}+e^{cx+c}+1\right)}{c^2} & \text{if } 0 < x \leq 1, \\ \frac{e^{-c}\left(e^{c}-1\right)^2}{c^2} & \text{if } x > 1, \end{cases} \quad (+\text{const}),$$

$$g_C(x,c) = \int k(x)e^{-xc}x\,\mathrm{d}x = \begin{cases} 0 & \text{if } x < -1 \\ \frac{e^{-cx}\left(-c^2x(x+1)-c(e^{cx+c}+2x+1)+2(e^{cx+c}-1)\right)}{c^3} & \text{if } -1 \leq x \leq 0, \\ \frac{e^{-cx}\left(c^2(x-1)x-c(e^{cx+c}-2x+1)+2(e^{cx-c}+1-2e^{cx})\right)}{c^3} & \text{if } 0 < x \leq 1, \\ -\frac{e^{-c}\left(e^{c}-1\right)(e^{c})c+c-2e^{c}+2}{c^3} & \text{if } x > 1, \end{cases} \quad (+\text{const}),$$

We the results of a numerical simulation using the Box kernel in Figure C.3.

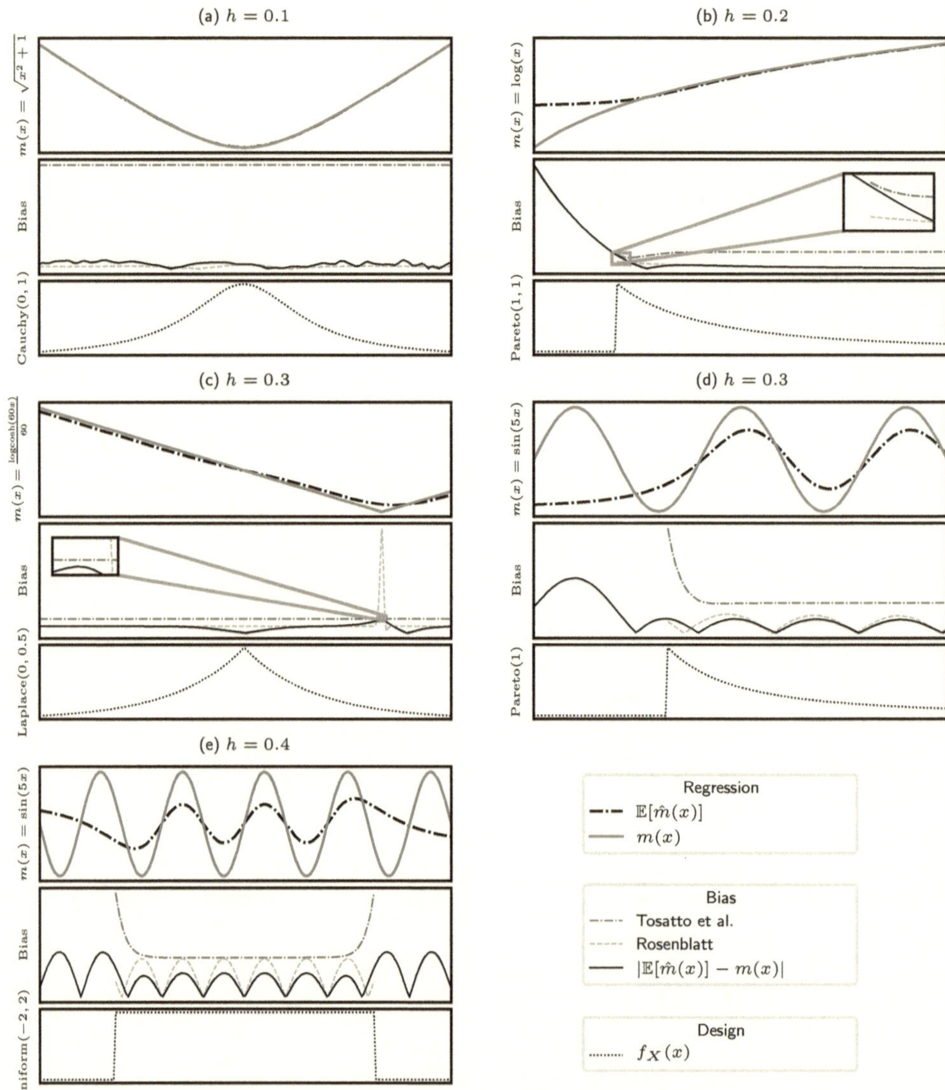

Figure C.1.: *Same experiment as in Figure 6.1, with Gaussian Kernels.*

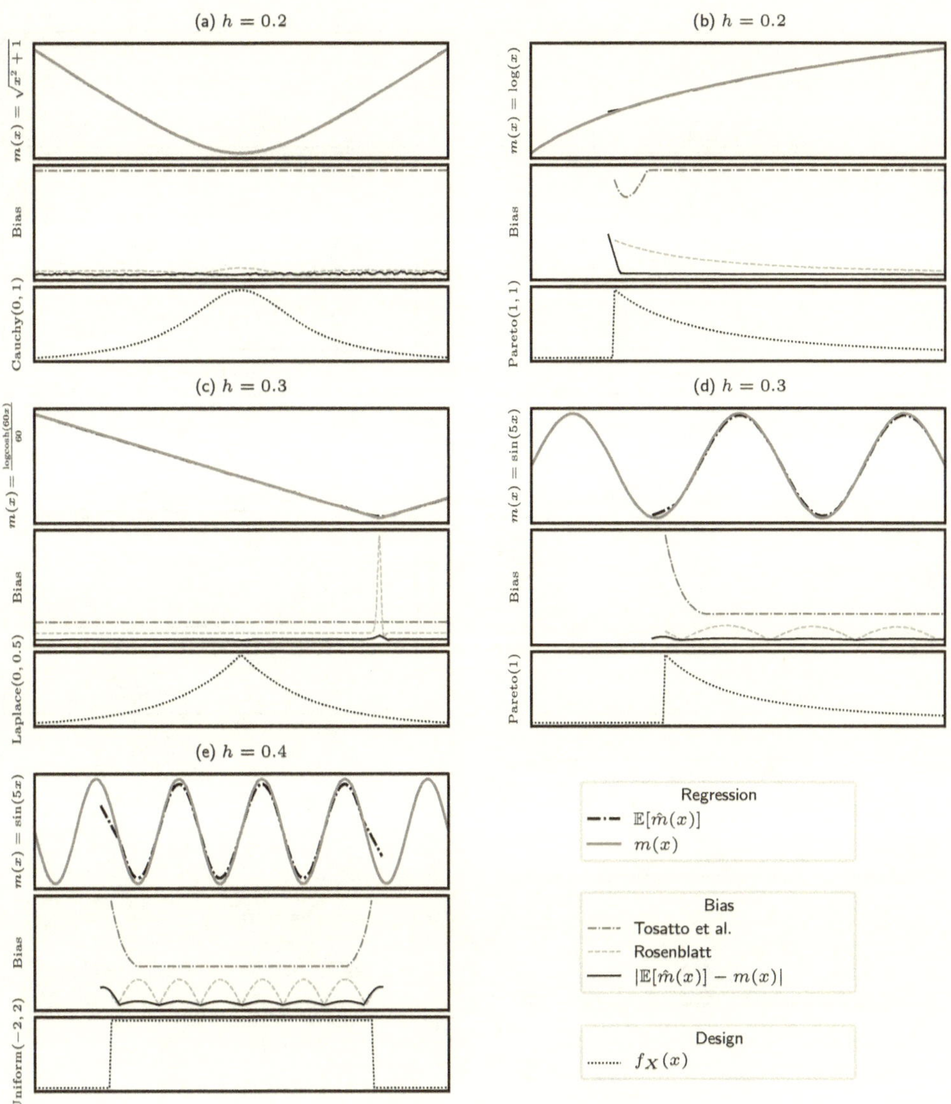

Figure C.2.: *Same experiment as in Figure 6.1, with Box Kernel.*

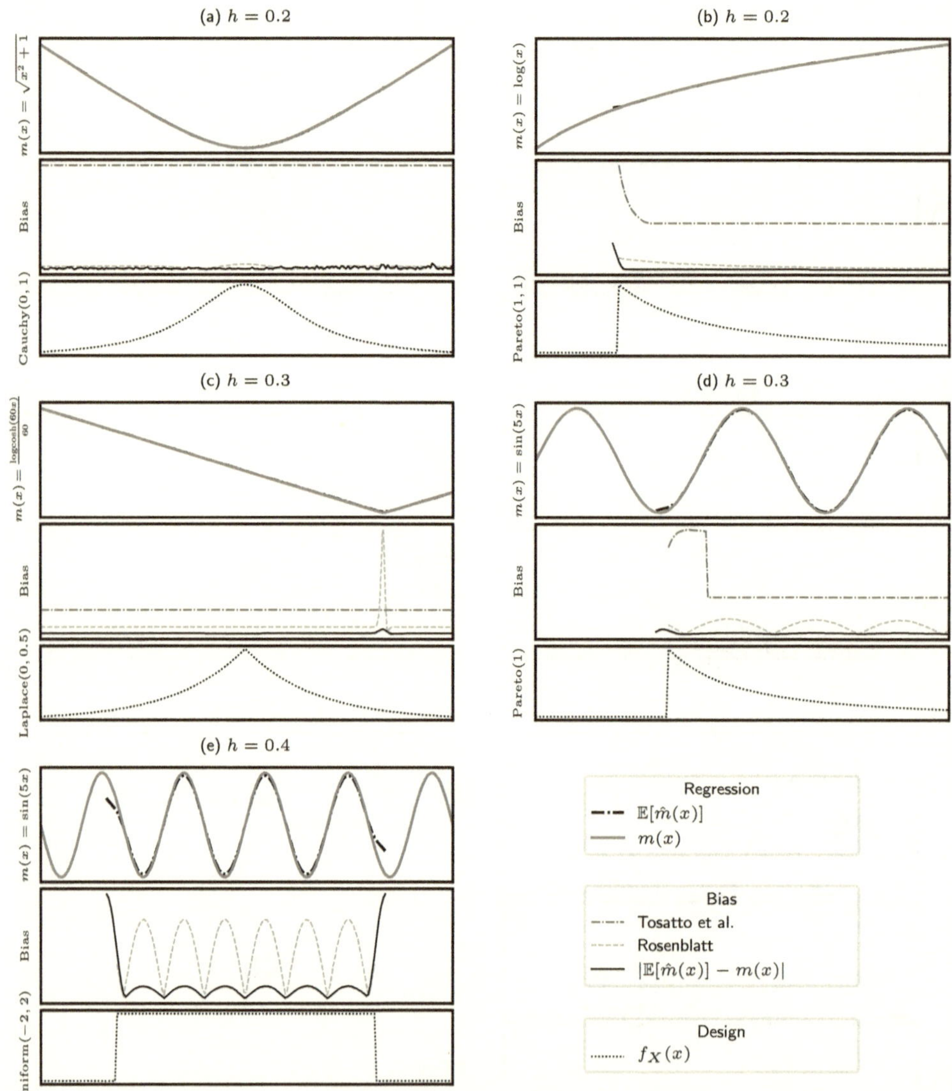

Figure C.3.: *Same experiment as in Figure 6.1, with Triangular Kernel. Note that this particular kernel, does not have finite bound for $L_f = 0$.*

D. Exploration Driven By an Optimistic Bellman Equation

This supplement provides details on all the proofs presented in Chapter 7 and shows experimental details that have been left out of the main paper due to space constraints. We refer to equations and definitions that are presented in the main paper with their assigned numbers.

First, we introduce additional notation for the classic Bellman operator and our optimistic version. We show how to derive the optimistic Bellman equation from an entropic regularized version of the bellman equation defined on a Q-value ensemble. Subsequently, we show how to interpret the optimistic Bellman equation under an intrinsic motivation perspective; we will show how the exploration bonus is related to the central moments of the approximations made. We then analyze two algorithms: Optimistic Value Iteration (OVI), which is presented mainly for theoretical reasons, introducing the optimistic Bellman operator and its properties, and optimistic Q-learning (OQL). For both algorithms we provide convergence proofs. We also show that the exploration bonus decreases according to the learning rate and state visits. Finally, we report the hyper-parameters used in our experimental setting.

D.1. Preliminaries

This section presents the mathematical notation used in the proofs.

Definition 8 (Bellman operator). *We define the Bellman operator* $\mathcal{T} : (\mathcal{S} \times \mathcal{A} \to \mathbb{R}) \to (\mathcal{S} \times \mathcal{A} \to \mathbb{R})$ *as:*

$$(\mathcal{T}Q)(s,a) = \overline{R}(s,a) + \gamma \int P(s'|s,a) \max_{a'} Q(s',a')\,\mathrm{d}s' \qquad \forall s,a \in \mathcal{S} \times \mathcal{A}$$

for each $Q : \mathcal{S} \times \mathcal{A} \to \mathbb{R}$.

Definition 9 (Optimistic Bellman operator). *We define the optimistic Bellman operator* $\mathring{\mathcal{T}}_{\eta}^{M} : (\mathcal{S} \times \mathcal{A} \to \mathbb{R})^{M} \to (\mathcal{S} \times \mathcal{A} \to \mathbb{R})^{M}$ *as:*

$$(\mathring{\mathcal{T}}_{\eta}Q)_i(s,a) = \overline{R}(s,a) + \eta^{-1} \log \sum_{m=1}^{M} \frac{e^{\eta\gamma \int P(s'|s,a) \max_{a'} Q_m(s',a')\,\mathrm{d}s'}}{M} \qquad \forall s,a,i \in \mathcal{S} \times \mathcal{A} \times \{1,\ldots,M\}$$

for each $Q : \mathcal{S} \times \mathcal{A} \to \mathbb{R}$.

Definition 10 (Optimal Q). *We define the optimal* Q^* *as the solution:*

$$\mathcal{T}Q^* = Q^*.$$

We know from dynamic programming that Q^* *exists and it is unique.*

Definition 11 (Optimistic Value Iteration (OVI)). *Let $Q = \{Q_i\}_{i=1}^{M}$ be an arbitrary set of Q-functions, let $\mathring{\mathcal{T}}_{\eta}^{N} Q$ denote the application of $\mathring{\mathcal{T}}_{\eta}$ N times on Q, let OVI be the procedure that computes $\tilde{Q} = \mathring{\mathcal{T}}_{\eta}^{N} Q$.*

D.2. Derivation of the Optimistic Bellman Equation

In this section, we derive the optimistic Bellman equation (OBE) from an entropic-regularized version of the bellman equation. The entropic-regularization is defined on a set of Q-value function.

Theorem 7. *Consider the problem:*

$$Q_i(s,a) = \max_{b(s,a)\in\mathcal{P}^M} f\big(s,a;b(s,a)\big) - \tfrac{1}{\eta}D_{\mathrm{KL}}\big(b(s,a)\|u\big)$$

$$s.t.\ \textstyle\sum_{m=1}^{M} b_m(s,a) = 1$$

$$\forall s,a,i \in \mathcal{S} \times \mathcal{A} \times \{1,\dots,M\}$$

where $f(a,s;p) = R(s,a) + \gamma\sum_m b_m(s,a)V_m'(s,a)$, $V_m'(s,a) = \sum_{s'} P(s'|s,a)\max_{a'} Q_m(s',a')$, $u_m = 1/M$, $D_{KL}(b(s,a)\|u)$ is the Kullback-Leibler divergence between the belief $b(s,a)$ and the uniform distribution u. From Problem 1, we can derive the optimistic Bellman equation (OBE)

$$Q_i(s,a) = \overline{R}(s,a) + \frac{1}{\eta}\log\frac{\sum_m e^{\eta\left(\gamma\sum_{s'} P(s'|s,a)\max_{a'} Q_m(s',a')\right)}}{M} \quad \forall i \in \{1,\dots,M\}.$$

Proof. Let L_i be the Lagrangian of the ith problem:

$$L_i(s,a) = \overline{R}(s,a)+\gamma\sum_m\left(\sum_{s'} P(s'|s,a)\max_{a'} Q_m(s',a')\right)b_m(s,a)-\frac{1}{\eta}\sum_m b_m(s,a)\log\frac{b_m(s,a)}{M^{-1}}+\lambda(\sum_m b_m(s,a)-1)$$

$$\tag{D.1}$$

To find the maximum of the Lagrangian, set the partial derivatives to zero. First, w.r.t. p_m:

$$\partial_{b_m(s,a)} L_i(s,a) \;=\; \gamma\sum_{s'} P(s'|s,a)\max_{a'} Q_m(s',a') - \frac{1}{\eta}\log\frac{b_m(s,a)}{M^{-1}} - \frac{M^{-1}}{\eta} + \lambda = 0$$

$$\implies \frac{1}{\eta}\log\frac{b_m(s,a)}{M^{-1}} \;=\; \gamma\sum_{s'} P(s'|s,a)\max_{a'} Q_m(s',a') - \frac{M^{-1}}{\eta} + \lambda$$

$$\implies b_m(s,a) \;=\; M^{-1}e^{\eta\left(\gamma\sum_{s'} P(s'|s,a)\max_{a'} Q_m(s',a')-\frac{M^{-1}}{\eta}+\lambda\right)}.$$

$$\tag{D.2}$$

Next, set partial derivative w.r.t. λ to zero:

$$\partial_\lambda L_i(s,a) \;=\; \sum_m b_m(s,a) - 1 = 0$$

$$\implies 1 \;=\; \sum_m M^{-1}e^{\eta\left(\gamma\sum_{s'} P(s'|s,a)\max_{a'} Q_m(s',a')-\frac{M^{-1}}{\eta}+\lambda\right)}$$

111

$$\implies e^{-\lambda\eta} \;=\; \sum_m M^{-1} e^{\eta\left(\gamma\sum_{s'} P(s'|s,a)\max_{a'} Q_m(s',a') - \frac{M^{-1}}{\eta}\right)}$$

$$\implies \lambda \;=\; -\frac{1}{\eta}\log\sum_m M^{-1} e^{\eta\left(\gamma\sum_{s'} P(s'|s,a)\max_{a'} Q_m(s',a') - \frac{M^{-1}}{\eta}\right)}. \tag{D.3}$$

Let's substitute (D.3) into (D.2):

$$b_m(s,a) \;=\; \frac{M^{-1} e^{\eta\left(\gamma\sum_{s'} P(s'|s,a)\max_{a'} Q_m(s',a') - \frac{M^{-1}}{\eta}\right)}}{\sum_{m'} M^{-1} e^{\eta\left(\gamma\sum_{s'} P(s'|s,a)\max_{a'} Q_{m'}(s',a') - \frac{M^{-1}}{\eta}\right)}}$$

$$\;=\; \frac{e^{\eta\left(\gamma\sum_{s'} P(s'|s,a)\max_{a'} Q_m(s',a')\right)}}{\sum_{m'} e^{\eta\left(\gamma\sum_{s'} P(s'|s,a)\max_{a'} Q_{m'}(s',a')\right)}}.$$

Now that we have solved for the Lagrangian multipliers, substitute $b_m(s,a)$ into Equation (D.1):

$$L_i(s,a) = \overline{R}(s,a) + \gamma\sum_m\left(\sum_{s'} P(s'|s,a)\max_{a'} Q_m(s',a')\right) b_m(s,a)(s,a) - \frac{1}{\eta}\sum_m b_m(s,a)\log\frac{b_m(s,a)}{u_m} \tag{D.4}$$

to get

$$L_i(s,a) = \overline{R}(s,a) + \gamma\sum_m\left(\sum_{s'} P(s'|s,a)\max_{a'} Q_m(s',a')\right) b_m(s,a)$$

$$-\frac{1}{\eta}\sum_m b_m(s,a)\log\frac{e^{\eta\left(\gamma\sum_{s'} P(s'|s,a)\max_{a'} Q_m(s',a')\right)}}{\sum_{m'} e^{\eta\left(\gamma\sum_{s'} P(s'|s,a)\max_{a'} Q_{m'}(s',a')\right)}} - \frac{\log M}{\eta}$$

$$= \overline{R}(s,a) + \gamma\sum_m\left(\sum_{s'} P(s'|s,a)\max_{a'} Q_m(s',a')\right) p_m(s,a)$$

$$-\frac{1}{\eta}\sum_m b_m(s,a)\left(\log e^{\eta\left(\gamma\sum_{s'} P(s'|s,a)\max_{a'} Q_m(s',a')\right)}\right.$$

$$\left. -\log\sum_{m'} e^{\eta\left(\gamma\sum_{s'} P(s'|s,a)\max_{a'} Q_{m'}(s',a')\right)}\right) - \frac{\log M}{\eta}$$

$$= \overline{R}(s,a) + \frac{1}{\eta}\sum_m b_m(s,a)\log\frac{\sum_{m'} e^{\eta\left(\gamma\sum_{s'} P(s'|s,a)\max_{a'} Q_{m'}(s',a')\right)}}{M}$$

$$= \overline{R}(s,a) + \frac{1}{\eta}\log\frac{\sum_m e^{\eta\left(\gamma\sum_{s'} P(s'|s,a)\max_{a'} Q_m(s',a')\right)}}{M} \tag{D.5}$$

Therefore (D.5) implies:

$$Q_i(s,a) = \overline{R}(s,a) + \frac{1}{\eta}\log\frac{\sum_m e^{\eta\left(\gamma\sum_{s'} P(s'|s,a)\max_{a'} Q_m(s',a')\right)}}{M} \qquad \forall i \in \{1,\dots,M\}.$$

$\square$

D.3. Exploration Bonus

We can show that OBE can be seen as an intrinsic motivation technique. More in detail, is it possible to see, that the optimistic over-estimation of the Q-ensembles is equivalent to an unbiased estimation of the ensemble plus a positive term which is connected with the uncertainty between values of the ensemble. We consider the unbiased estimate of the value of the next state

$$\overline{q}'(s, a) = \sum_{m=1}^{M} \frac{Q'_m(s, a)}{M}.$$

(D.6)

The exploration bonus can then be expressed as

$$
\begin{aligned}
U(s, a) &= \frac{1}{\eta} \log \left(\sum_{m=1}^{M} \frac{e^{\eta\gamma V'_m(s,a)}}{M} \right) - \overline{Q}'(s, a) \\
&= \frac{1}{\eta} \left(\log \left(\sum_{m=1}^{M} \frac{e^{\eta\gamma Q'_m(s,a)}}{M} \right) - \eta \overline{Q}'(s, a) \right) \\
&= \frac{1}{\eta} \log \left(\sum_{m=1}^{M} \frac{e^{\eta\gamma Q'_m(s,a)}}{M} e^{-\eta \overline{Q}'(s,a)} \right) \\
&= \frac{1}{\eta} \log \sum_{m=1}^{M} \frac{e^{\eta\gamma(Q'_m(s,a) - \overline{Q}'(s,a))}}{M}.
\end{aligned}
$$

(D.7)

Note that the average over the exponential function is the sample moment generating function. Thus, Equation (D.7) can be rephrased as

$$U(s, a) = \lim_{N \to +\infty} \frac{1}{\eta} \log \left[1 + \sum_{n=2}^{N} \frac{(\eta\gamma)^n}{n!} \mathcal{M}_n(s, a) \right].$$

where

$$
\begin{aligned}
\mathcal{M}_n(s, a) &= M^{-1} \sum_{m=1}^{M} \left[\left(V'_m(s, a) - \overline{V}(s, a) \right)^n \right] \\
&= \eta\gamma \mathcal{M}_2(s, a) + O(\eta^2)
\end{aligned}
$$

(D.8)

denotes the nth sample central moment. Note that the further simplification come from the expansion w.r.t. η.

D.4. Convergence of Value Iteration with the Optimistic Bellman Equation

In this Section, we show that value iteration using the optimistic Bellman equation (OBE), which we call optimistic value iteration (OVI), converges. First, we show that the fixed point of value iteration with OBE is identical to the fixed point of the classic Bellman equation (BE). Next, we show the max-norm contractivity of the optimistic Bellman operator. Finally, we use the fixed point and max-norm contractivity results to show that value iteration with OBE converges.

Lemma 1 (Fixed point of OBE). *If $Q_i = Q^*$ then $(\mathring{\mathcal{T}}_\eta Q)_i = Q^* \quad \forall i \in \{1, \ldots M\}$ where Q^* is the unique fixed point of the classic BE .*

Proof.

$$
\begin{aligned}
(\tilde{T}^* Q^*)(s, a) &= \overline{R}(s, a) + \frac{1}{\eta} \log \frac{1}{M} \sum_{i=1}^{M} e^{\eta\gamma \int P(s'|s,a) \max_{a'} Q^*(s',a') \, ds'} \\
&= \overline{R}(s, a) + \frac{1}{\eta} \log e^{\eta\gamma \int P(s'|s,a) \max_{a'} Q^*(s',a') \, ds'} \\
&= \overline{R}(s, a) + \gamma \int P(s'|s, a) \max_{a'} Q^*(s', a') \, ds' \\
&= (T^* Q^*)(s, a) \\
&= Q^*(s, a)
\end{aligned}
$$

$\square$

Lemma 2 (Max-Norm contractivity of the optimistic Bellman operator). *Given $\{Q_{1,k}\}_{k=1}^{M}$, $\{Q_{2,k}\}_{k=1}^{M}$, and $\delta > 0$ such that*

$$
\|Q_{1,k} - Q_{2,k}\|_\infty \leq \delta \qquad \forall k \in \{1, \ldots, M\}
$$

implies that:

$$
\|(\mathring{\mathcal{T}}_\eta Q_1)_k - (\mathring{\mathcal{T}}_\eta Q_2)_k\|_\infty \leq \gamma\delta \qquad \forall k \in \{1, \ldots, M\}.
$$

Proof.

$$
\begin{aligned}
(\mathring{\mathcal{T}}_\eta Q_1)_k(s, a) - (\mathring{\mathcal{T}}_\eta Q_2)_k(s, a) &= \overline{R}(s, a) + \frac{1}{\eta} \log \frac{1}{M} \sum_{i=1}^{M} e^{\eta\gamma \int P(s'|s,a) \max_{a'} Q_{1,i}(s',a') \, ds'} \\
&\quad -\overline{R}(s, a) - \frac{1}{\eta} \log \frac{1}{M} \sum_{i=1}^{M} e^{\eta\gamma \int P(s'|s,a) \max_{a'} Q_{2,i}(s',a') \, ds'} \\
&= \frac{1}{\eta} \log \frac{1}{M} \sum_{i=1}^{M} e^{\eta\gamma \int P(s'|s,a)(\max_{a'} Q_{1,i}(s',a') - \max_{a'} Q_{2,i}(s',a')) \, ds'} \\
&\leq \frac{1}{\eta} \log \frac{1}{M} \sum_{i=1}^{M} e^{\eta\gamma \int P(s'|s,a)\delta \, ds'} \\
&= \gamma\delta
\end{aligned}
$$

$\square$

Theorem 8 (Convergence of OBE). *If*

$$
|Q_i(s, a) - Q^*(s, a)| \leq \epsilon \qquad s, a, i \in \mathcal{S} \times \mathcal{A} \times \{1, \ldots, M\} \tag{D.9}
$$

then

$$
|(\mathring{\mathcal{T}}_\eta Q)_i(s, a) - Q^*(s, a)| \leq \gamma\epsilon \qquad s, a, i \in \mathcal{S} \times \mathcal{A} \times \{1, \ldots, M\}. \tag{D.10}
$$

Note that this implies that given an initial set of $Q = \{Q_i\}_{i=1}^{M}$, $\lim_{N \to \infty} \mathring{\mathcal{T}}_\eta^N Q = Q^*$, therefore implies the convergence of OVI.

Proof.

$$\left|(\mathring{\mathcal{T}}_\eta Q)_i(s,a) - Q^*(s,a)\right| = \left|(\mathring{\mathcal{T}}_\eta * Q)_i(s,a) - \mathring{\mathcal{T}}_\eta Q^*(s,a)\right|$$
$$\leq \gamma\epsilon$$

$\square$

We note that OVI converges with the same rate as VI.

D.5. Optimistic Q-Learning

This section provides a more detailed proof of optimistic Q-learning (OQL) converging to $Q(s,a)^*$. We first show that the exploration bonus vanishes in OQL and then give the main proof of convergence.

Definition 12 (Optimistic Q-learning - *theoretical version*). [1]

$$\begin{cases} Q_{j,t+1}(s,a) = (1-\alpha_t)Q_{i,t}(s,a) + \alpha_t\left(r_t + \gamma\frac{1}{M}\sum_{j=1}^{M}\max_{a'} Q_{j,t}(s_{t+1},a')\right) & \text{if } s = s_t \wedge a = a_t \\ Q_{i,t+1}(s,a) = Q_{i,t}(s,a) & \text{otherwise} \end{cases}$$

Theorem 9 (Vanishing bonus for optimistic Q-learning). *Let's consider optimistic Q-Learning (OQL) described in Definition 12. Let's suppose to have a set of Q_i. Each entry in the table at time $t = 0$ has central moment $\mathcal{M}_{0,n} \in \mathbb{R}$. If we consider a specific entry (s,a), and a sequence of learning rate $\{\alpha_t\}_{t=0}^T$ where $\alpha_t \in [0,1]$, then*

$$\mathcal{M}_{T+1,n}(s,a) = \prod_{t=0}^{T}(1-\alpha_t)^n \mathcal{M}_{0,n}(s,a) \tag{D.11}$$

where $\mathcal{M}_{T+1,n}(s,a)$ is the nth central moment of the entry (s,a) at time $T+1$.

Proof. Please, note that we always update an entry (s,a) with the same value for all the M tables. We can refer to the sequence of updates to a single state-action pair (s,a) as $\{y_t\}_{t=0}^T$, where each value belongs to $\mathbb{R}$. Now, let's consider the process of updating of the entry s,a

$$Q_{t+1,i}(s,a) = (1-\alpha_t)Q_{t,i}(s,a) + \alpha_t y_t \qquad \forall i \in \{1,\ldots M\}. \tag{D.12}$$

We can write the central moments at time $t+1$ as

$$\begin{aligned} \mathcal{M}_{t+1,n}(s,a) &= M^{-1}\sum_{m}\left(Q_{t+1,m}(s,a) - M^{-1}\sum_{i}Q_{t+1,i}(s,a)\right)^n \\ &= M^{-1}\sum_{m}\left((1-\alpha_t)Q_{t,m}(s,a) + \alpha_t - M^{-1}\sum_{i}((1-\alpha_t y_t)Q_{t,i}(s,a) + \alpha_t)\right)^n \\ &= M^{-1}\sum_{m}\left((1-\alpha_t)Q_{t,m}(s,a) - M^{-1}\sum_{i}(1-\alpha_t y_t)Q_{t,i}(s,a)\right)^n \end{aligned}$$

---

[1] We use α_t as a shortcut for $\alpha_t(s,a)$.

$$
\begin{aligned}
&= (1-\alpha_t)^n M^{-1} \sum_m \Big(Q_{t,m}(s,a) - M^{-1} \sum_i Q_{t,i}(s,a) \Big)^n \\
&= (1-\alpha_t)^n \mathcal{M}_{t,n}(s,a)
\end{aligned}
$$

and therefore, unfolding the recursion,

$$
\mathcal{M}_{T+1,n}(s,a) = \prod_{t=0}^{T} (1-\alpha_t)^n \mathcal{M}_{0,n}(s,a)
$$

$\square$

Next, we prove the convergence of OQL. Our proof is based on the work of Melo (2001), which relies on the following theorem Jaakkola et al. (1994):

Theorem 10. *The random process $\{\Delta_t\}$ taking values in $\mathbb{R}$ and defined as:*

$$
\Delta_{t+1}(x) = (1-\alpha_t(x))\Delta_t(x) + \alpha_t F_t(x) \tag{D.13}
$$

converges to zero w.p. 1 under the following assumptions:

- $0 \leq \alpha_t \leq 1$, $\sum_t \alpha_t(x) = \infty$ and $\sum_t \alpha_t^2(x) < \infty$
- $\|\mathbb{E}[F_t(x)|\mathcal{F}_t]\|_W \leq \gamma\|\Delta_t\|_W$ with $0 \leq \gamma < 1$
- $\mathrm{Var}[F_t(x)|\mathcal{F}_t] \leq C(1 + \|\Delta_t\|_W^2)$ for $C > 0$.

We are now ready to prove the convergence of optimistic Q-learning.

Theorem 11 (Convergence of Optimistic Q-learning). *Let us consider the algorithm provided in Definition 12. Suppose that the rewards are bounded, and consider a learning rate $\alpha_t(s,a)$ which satisfies $0 \leq \alpha_t \leq 1$, $\sum_t \alpha_t(x) = \infty$ and $\sum_t \alpha_t^2(x) < \infty$. Suppose that each state action pair is visited infinitely many times, then, $\lim_{t \to \infty} Q_{i,t}(s,a) \to Q^*(s,a) \forall i \in \{1, \dots, M\}$ with probability 1. Then the algorithm converges.*

Proof. Let's consider the following stochastic process:

$$
Q_{i,t+1}(s,a) = (1-\alpha_t)Q_{i,t}(s,a) + \alpha_t \Big(r_t + \frac{1}{\eta} \log M^{-1} \sum_{j=1}^{M} e^{\gamma \max_{a'} Q_{j,t}(s_{t+1},a')} \Big)
$$

with $\alpha_t = \alpha_t(s,a)$ (for brevity), and coherent with the assumpions. Let's rename $\Delta_{i,t}(s,a) = Q_{i,t}(s,a) - Q^*(s,a)$ where $Q^*(s,a)$ is the fixed point of $\mathring{\mathcal{T}}_\eta$. We can now write:

$$
\Delta_{i,t+1}(s,a) = (1-\alpha_t)\Delta_{i,t}(s,a) + \alpha_t \Big(r_t + \frac{1}{\eta} \log M^{-1} \sum_{j=1}^{M} e^{\gamma \max_{a'} Q_{j,t}(s_{t+1},a')} - Q^*(s,a) \Big).
$$

We can rename $F_t(s,a) = r_t + \frac{1}{\eta} \log M^{-1} \sum_{j=1}^{M} e^{\gamma \max_{a'} Q_{j,t}(s_{t+1},a')} - Q^*(s,a)$ and observe that $\mathbb{E}[F_t(s,a)] = \mathring{\mathcal{T}}_\eta Q_t(s,a) - Q^*(s,a)$, and subsequently, thanks to the max-norm contraction of the optimistic Bellman operator,

$$
\|\mathbb{E}[F_t(s,a)]\|_\infty \leq \gamma\|Q_{i,t} - Q^*\|_\infty
$$

$$= \gamma \|\Delta_{i,t+1}(s,a)\|_\infty \qquad \forall i \in \{1,\dots,M\}.$$

Then, noticing $\mathrm{Var}[F_t(s,a)] = \mathrm{Var}[F_t(s,a) - Q^*] = \mathrm{Var}[r_t + \frac{1}{\eta} \log M^{-1} \sum_{j=1}^{M} e^{\gamma \max_{a'} Qj,t(s_{t+1},a')}]$ which, considering that the rewards are assumed to be bounded, leads to:

$$\exists C : \mathrm{Var}[F_t(s,a)] \le C(1 + \|\Delta_{i,t}\|) \qquad \forall i \in \{1,\dots,M\}. \tag{D.14}$$

Using Theorem 10 we can therefore say that $Q_t(s,a)$ converges to $Q^*(s,a)$ w.p. 1. $\qquad\square$

D.6. Details on the Experiments

In this section, we provide details on the exact form of bootstrapped Q-Learning and optimistic Q-learning update rules. Furthermore, we describe how hyper-parameters were chosen in the experiments and what kind of neural network structure was used in the experiments with neural network based function approximation.

We define Bootstrapped Q-Learning (BQL) with the following update rule

Definition 13 (Bootstrapped Q-learning). [2]

$$\begin{cases} Q_{i,t+1}(s,a) = (1 - \alpha_t)Q_{i,t}(s,a) + \alpha_t\left(r_t + \frac{1}{\eta}\log M^{-1} \sum_{j=1}^{M} e^{\gamma \max_{a'} Qj,t(s_{t+1},a')}\right) & \text{if } s = s_t \wedge a = a_t \\ Q_{i,t+1}(s,a) = Q_{i,t}(s,a) & \text{otherwise} \end{cases}$$

and OQL with

Definition 14 (Optimistic Q-learning - *empirical analysis version*).

$$\begin{cases} Q_{j,t+1}(s,a) = (1 - \alpha_t)Q_{i,t}(s,a) + \alpha_t\Big(r_t + \frac{1}{\eta}\log \frac{1}{M-1} \sum_{j=2}^{M} e^{\gamma \max_{a'} Q_{j,t}(s_{t+1},a')+Q_1(s_{t+1},a')} \\ \qquad\qquad -\gamma\frac{1}{M-1}\sum_{j=2}^{M} \max_{a'} Qj,t(s_{t+1},a')\Big) & \text{if } s = s_t \wedge a = a_t \wedge \\ Q_{j,t+1}(s,a) = (1 - \alpha_t)Q_{i,t}(s,a) \\ \qquad\qquad +\alpha_t\Big(r_t + \gamma\frac{1}{M-1}\sum_{j=2}^{M} \max_{a'} Q_{j,t}(s_{t+1},a')\Big) & \text{if } s = s_t \wedge a = a_t \wedge \\ Q_{i,t+1}(s,a) = Q_{i,t}(s,a) & \text{otherwise} \end{cases}$$

using the "explicit exploration" formulation, in order to enable the evaluation with unbiased Q values. The settings provided by Table D.1 are fixed for all the different environments.

---

[2] We use α_t as a shortcut for $\alpha_t(s,a)$.

Parameter	QL	OIQL	BQL	OQL
Number of approximators	1	1	10	10
Initialization	$\mathcal{N}(0,2)$	$R_{\max}$	$\mathcal{N}(0,2)$	$\mathcal{N}(0,2)$
Learning rate	0.15	0.15	0.15	0.15
ϵ-greedy	0.01	0.01	0.01	0.01

Table D.1.: Setting used for tabular algorithms.

Acrobot configuration. For Acrobot, we used a single-layer neural network as base component of the ensemble. The input of the neural newtork is 4-dimensional (and corresponds to the dimension of the state space), and the output has 2 dimensions, corresponding to the two possible actions. Table D.4 shows the fixed hyper-parameters for Acrobot. Additionally, we performed a grid search over the number of neurons in the hidden layer, and for the "bootstrapped mask" (see Table D.2). We measured both the mean return averaged over all episodes denoted by "avg" (which should give an idea about how fast an algorithm can improve performance w.r.t. the number of samples), and also the mean of the final performance. For both ODQN and BDQN we selected the hyper-parameters corresponding to the best average performance of BDQN yielding yielding 100 neurons and a bootstrap mask of 0.5 for the final performance evaluation shown in the main paper. For ODQN we use $\chi = 0.25$ and $\iota_{\max} = 1$.

Neurons	Bootstrapped Mask	BDQN avg	BDQN final	ODQN avg	ODQN final
100	0.5	-116.96^*	-84.63	-115.25	-86.04
100	1.0	-129.62	-86.91	-129.06	-95.60
150	0.5	-123.21	-85.04	-122.10	-80.38
150	1.0	-136.25	-89.50	-138.38	-83.62
200	0.5	-123.47	-87.89	-125.72	-84.41
200	1.0	-143.05	-87.70	-148.64	-81.91
300	0.5	-129.26	-82.60	-131.24	-81.12
300	1.0	-150.10	-83.98	-151.14	-86.30
400	0.5	-133.01	-81.43	-135.30	-83.82
400	1.0	-154.70	-87.58	-158.38	-86.55

Table D.2.: Tested hyper-parameters "Neurons" and "Bootstrapped Mask" for Acrobot with corresponding evaluations.

Taxi configuration. For Taxi, we decided to encode the state as a 2-dimensional grid, selecting only the position of the agent to 1 and the rest to zero, and additionally we provide a one-dimensional vector of length 3 providing the information about which flags where collected. We decided to use a shared convolutional layer with kernel of 2 and stride 1, in order to process the vector and reduce the dimension. Above the convolutional layer, we apply a different hidden layer for each component of the ensemble. The output of each component is 4-dimensional, corresponding to the four possible actions. Most of the parameters are chosen without any optimization (see Table D.4), except for the number of neurons in the hidden layer, and for the "bootstrapped mask". We performed a grid search over these two parameters, measuring both the mean return averaged over all episodes and also the mean of the final performance, similarly to the Acrobot evaluation (please, see Table D.3). For both ODQN and BDQN, we selected the hyper-parameters corresponding to the best average performance of BDQN yielding 200 neurons and a bootstrap mask of 0.5 for the final performance evaluation shown in the main paper.

Neurons	Bootstrapped Mask	BDQN avg	BDQN final	ODQN avg	ODQN final
200	0.5	9.51*	12.05	**10.05**	**14.25**
200	1.0	6.16	7.80	**8.63**	**12.00**
300	0.5	9.14	**10.95**	**9.63**	9.65
300	1.0	**9.36**	9.65	9.27	**10.55**
400	0.5	7.21	8.4	**10.30**	**13.50**
400	1.0	6.87	9.2	**9.42**	**11.25**

Table D.3.: Tested hyper-parameters "Neurons" and "Bootstrapped Mask" for Taxi with corresponding evaluations.

Parameter	Acrobot	Taxi
Number of approximators	10	10
Shared conv. layer	no	yes
Number of layers	1	1
Number of neurons	100	200
Activation function	relu	relu
Initialization	Glorot Uniform	Glorot Uniform
Loss	MSE	Huber Loss
Optimization	Adam	RMSProp
Learning rate	0.001	0.00075
Decay (only RMSProp)	none	0.95
Batch size	32	100
Max replay-memory size	5000	100000
Target update frequency	600	100
p-mask	0.5	0.5
ϵ-greedy (training)	0.0	0.05
ϵ-greedy (evaluation)	0.0	0.0
Evaluation frequency	3000	5000
Total training steps	250000	400000

Table D.4.: Common hyper-parameters for BDQN and ODQN.

E. Publication List

E.1. Journal Papers

Tosatto, S.; Carvalho, J.; Peters, J. (Under Revision). Batch Reinforcement Learning with a Nonparametric Off-Policy Policy Gradient, *Transaction of Pattern Analysis and Machine Intelligence (TPAMI)*.

Tosatto, S.; Chalvatzaki, G.; Peters, J. (Under Revision). Contextual Latent-Movements Off-Policy Optimization for Robotic Manipulation Skills, *IEEE Robotics and Automation Letters (RA-L)*.

Tosatto, S.; Akrour, R.; Peters, J. (2021). An Upper Bound on the Bias of Nadaraya-Watson Kernel Regression under Lipschitz Assumptions *MDPI Stats*.

E.2. Conference Papers

Tosatto, S.; Pirotta, M.; D'Eramo, C; Restelli, M. (2017). Boosted Fitted Q-Iteration, *Proceedings of the International Conference of Machine Learning (ICML)*.

Rueckert, E.; Nakatenus, M.; **Tosatto, S.**; Peters, J. (2017). Learning Inverse Dynamics Models in O(n) time with LSTM networks, *Proceedings of the IEEE International Conference on Humanoid Robots (HUMANOIDS)*.

Tosatto, S.; D'Eramo, C.; Pajarinen, J.; Restelli, M.; Peters, J. (2019). Exploration Driven By an Optimistic Bellman Equation, *Proceedings of the International Joint Conference on Neural Networks (IJCNN)*.

Tosatto, S.; Carvalho, J.; Abdulsamad, H.; Peters, J. (2020). A Nonparametric Off-Policy Policy Gradient, *Proceedings of the International Conference on Artificial Intelligence and Statistics (AISTATS)*.

E.3. Workshops

Tosatto, S.; Carvalho, J.; Abdulsamad, H.; Peters, J. (2020). A Nonparametric Off-Policy Policy Gradient, *Workshop on Reinforcement Learning and Decision Making (RLDM)*.

F. Curriculum Vitae

Research Interests

Reinforcement Learning: (approximate) dynamic programming; policy gradient techniques; model based; exploration/exploitation tradeoff; risk awareness (in MDPs); optimal control;
Machine Learning: linear models; deep models; mixture models; kernel methods; boosting; dimensionality reduction;
Robotics: Human-robot interatcion, movement primitives, robot-learning.

Education

May 2017 - 2021	Technische Universität Darmstadt, Germany, **Ph.D. Student**
Sept. 2015 - Feb. 2016	ERASMUS Johannes Kepler Universität, Linz, Austria
Sept. 2013 - Feb. 2017	Polytechnic University of Milan, Italy, **Master in Software Engineering** (110/110
Sept. 2009 - June 2012	Polytechnic University of Milan, Italy, **Bachelor in Software Engineering** (96/110

Additional Experience

2017   SPP 1527 Summer School on Robotics & Autonomous Learning

Programming Skills

Programming Languages: Python; Java; C#;
Libraries/Frameworks/Revision Controls: Numpy; Tensorflow; PyTorch; Scipy; Scikit-Learn; CVXPY; Cython; V-REP/Coppelia; ROS (and PyROS); Git.

Natural Languages

Italian (Mother tongue), English (C1), German (B1).

Student Supervision

2020	Bachelor book	Dimensionality Reduction of Movement Primitives in Parameter Space Stadtmüller, J.; Tosatto, S.; Peters, J.
2019	Master book	A Nonparametric Off-Policy Policy Gradient Carvalho, J.; Tosatto, S.; Peters, J.
2019	Master book	Improving Sample-Efficiency with a Model-Based Deterministic Policy Gradient Saoud, H.; Tosatto, S.; Peters, J.
2018	Master book	Boosted Deep Q-Network Tschirner, J.; Tosatto, S.; Peters, J.

Reviewing

2020	International Conference of Machine Learning (ICML)
2019, 2020	IEEE Robotics and Automation Letters (RA-L)
2018	IEEE International Conference on Humanoid Robots (HUMANOIDS)
2018	International Conference on Intelligent Robots and Systems (IROS)

References

Prof. Jan Peters, TU Darmstadt and Max Planck Institute for Machine Intelligence, mail@jan-peters.net, +49 6151-16-25374

Prof. Marcello Restelli, Polythecnic University of Milan, marcello.restelli@polimi.it, +39 02-2399-4015

Prof. Joni Pajarinen, Aalto University, joni.pajarinen@aalto.fi, +35 850 3094771

Prof. Elmar Rückert, Universität zu Lübeck, rueckert@rob.uni-luebeck.de, +49 451 31015209